Acknowledgements

Research contractors

This report is the result of a research project carried out by FaberMaunsell in partnership with the University of Surrey, under contract to CIRIA.

Authors

J M C Cadei

T J Stratford

L C Hollaway

W G Duckett

CIRIA manager

The project was developed and managed by Arna Peric-Matthews at CIRIA.

Steering group

Following CIRIA's usual practice, the research project was guided by a steering group, whose support and valuable contributions are gratefully acknowledged.

Chairman   Mr Bob Cather   Ove Arup and Partners

Deputy chairman   Dr Andrew McLeish   WS Atkins

Members

Mr Brian Bell	Network Rail
Dr John Clarke	The Concrete Society
Mr Peter Dennis	Balvac Whitley Moran
Dr Adrian Dier	MSL Engineering
Mr Neil Farmer	Tony Gee and Partners
Dr Simon Frost	AEA Technology
Dr Sam Luke	Mouchel
Mr Mike Langdon	MBT Feb
Dr Mick Mahon	Toray Europe
Mr Jim Moriarty	London Underground
Dr Stuart Moy	University of Southampton
Mr John Powell	British Waterways

Corresponding members

Mr Richard Barnes	Cranfield University
Dr Steve Denton	Parsons Brinckerhoff
Dr Holger Garden	Taywood Engineering
Mr Terry Girdler	English Heritage
Dr Paul Hill	DML Composites
Mr Michael Johnston	British Energy Generation (UK)
Ms Sarah Kaethner	Ove Arup and Partners
Mr Eamon McDonnell	Weber & Broutin UK
Dr Toby Mottram	University of Warwick

Project funders   CIRIA and the authors gratefully acknowledge the support of the following funding organisations and the technical help and advice provided by the members of the steering group. Contributions do not imply that individual funders necessarily endorse all views expressed in published outputs.

DTI Partners in Innovation
Railtrack
Weber SBD
British Waterways
British Energy
MBT
Toray Europe
London Underground

Contents

Figures

Tables

Glossary

Ablation	The removal of a substance by a destructive process such as pyrolysis
Adherends	The components bonded together by an adhesive
Adhesive	A polymeric material that is capable of bonding two materials together by surface attachment
Adhesive system	The combination of the adhesive layer plus any primer layer
AFRP, CFRP, GFRP	Aramid, carbon and glass fibre-reinforced-polymer composites
Anisotropic	Having different mechanical properties in different directions
Aramid fibre	A high-strength, high-stiffness aromatic polyamide fibre. Also known as poly-aramid (trade names Kevlar and Twaron)
Bond	The adhesion of one surface to another, with the use of an adhesive or other bonding agent
Brick jack arch	A form of construction in which short-span brick barrel vaults span between cast iron or (less commonly) wrought iron or steel beams. Usually built with tie-rods between adjacent beams. Much used from the late 18th century to the early 20th century
Brittle	The inability of a material to sustain imposed load or deformation after reaching a critical stage of the loading cycle. This property is associated with non-ductile materials
Carbon fibre	High-modulus, high-strength fibres produced by the pyrolysis of organic precursor fibres such as rayon, polyacrylonitrile (PAN) or pitch in an inert atmosphere. The term is often used interchangeably with graphite. However, carbon fibres and graphite fibres differ in the temperature at which the fibres are made and heat-treated, and in the carbon content
Cast iron	A mouldable alloy of iron and about 1.8–5 per cent of carbon, also containing small amounts of sulphur, phosphorous, silicon (and often manganese)
Composite	A material that combines fibre and a binding matrix to obtain specific performance properties. Neither element merges completely with the other. Advanced polymer composites use only continuous, orientated fibres in a polymer matrix
Consolidation	A process step that compresses fibre and matrix to remove excess resin, reduce voids and achieve a particular density
Cure	To change irreversibly the molecular structure and physical properties of a thermosetting resin to a more stable condition with the desired properties. The chemical reaction may be assisted by heat and catalysts either singly or in combination, with or without pressure
Debond	A non-adherent or unbonded region in an adhesive
DMTA	Dynamic mechanical thermal analysis test method for determining the glass transition temperature of a polymer
DSC	Differential scanning calorimetry test method for determining the glass transition temperature of a polymer
Ductile cast iron	Also known as nodular or spheroidal cast iron. A form of cast iron in which the graphite inclusions are spherical rather than flaky in shape, conferring on the material similar tensile strength and ductility properties to steel
E-glass fibre	A commonly used grade of structural glass fibre

Epoxy resin	Resins of widely different chemical structure characterised by the reaction of the epoxy group to form a cross-linked hard resin
Fabric	A material constructed of interlaced yarns, fibres or filaments, usually planar
Fibre reinforced polymer/FRP	See **Composite**
Galvanic corrosion	Corrosion resulting from the intimate contact of two conducting materials that are separated on the electropotential series. Also referred to as bimetallic corrosion
Glass fibre	A fibre spun from an inorganic product of fusion that has cooled to a rigid condition without crystallising
Glass transition	The approximate mid-point of the temperature range over which an amorphous material changes from a brittle, vitreous state to a plastic, viscous state
Grey/flake cast iron	A form of cast iron that has been allowed to cool slowly in which most of the carbon is deposited as graphite flakes. It is brittle, and relatively weak in tension but fairly strong in compression
Hand lay-up	A process in which resin and reinforcement are applied manually either to a mould or to a working surface in successive layers
Heat distortion temperature (HDT)	Temperature at which a test bar deflects a certain amount under a specified temperature curve and load
Hygroscopic	A material that readily absorbs water
Hygrothermal effects	Moisture content and temperature conditions affecting the performance of a material
Laminate	A composite material consisting of one or more layers of fibre, impregnated with a resin system and cured
Lap-shear test	A test undertaken to determine the adhesion and adhesive properties of an adhesive, in which two pieces of substrate are bonded together and the test piece loaded in tension such that the overlapping area is loaded in shear
Load-relief jacking	A method for transferring a portion of the dead load stress from the existing structure into the strengthening by jacking the structure before the strengthening material is applied, and then removing the jacks after the strengthening has cured
Matrix	In reinforced polymers the matrix is the resin binder and any filler, which impregnates and consolidates the fibre reinforcement so as to produce a composite engineering material
Nodular cast iron	See **Ductile cast iron**
Peel ply	Sacrificial surface layer of reinforcing material included in a laminate that is removed from the cured laminate prior to bonding operations so as to leave a clean surface suitable for bonding
Peel stress	Tensile stress acting normal to an adhesive interface tending to pull the strengthening material away from the substrate
Polyesters	Thermosetting resins produced by dissolving unsaturated alkyd resins in a vinyl-type active monomer, such as styrene, methyl styrene and diallyl phthalate. Cure is effected through vinyl polymerisation using peroxide catalysts and promoters, or heat, to accelerate the reaction. The resins are usually furnished in solution form, but powdered solids are also available
Pot life	The length of time in which a catalysed thermosetting resin retains sufficiently low viscosity for processing
Preformed FRP	Composite strengthening material that is formed and cured into the required shape before it is bonded to the structure requiring strengthening

Prepreg (pre-impregnated fibre)	Resin-impregnated cloth, mat or filaments in flat form that can be stored for later use in moulds or wet lay-up. The resin is often partially cured to a tack-free state
Primer	Material used to protect or regularise a surface before the adhesive is applied, to improve adhesion and/or improve the durability or stabilise/protect the substrate
Pultrusion	A semi-continuous process for the manufacture of structural shapes of constant section. After being passed through a resin tank, rovings and other reinforcements are drawn through a heated die at a constant velocity and emerge as a cured finished cross-section
Putty	A filler that is used to repair and fill irregularities in a surface to which strengthening material will later be applied. The putty must be compatible with the strengthening system used, and will normally be polymeric
Resin	A polymer that may exist in solid, semi-solid, or liquid state. Thermosetting resins are capable of repeated softening by an increase in temperature, and hardening by a decrease in temperature. Thermoplastic resins are cured by heat and pressure or with a catalyst into an infusible and insoluble material. Once cured, a thermoset cannot be returned to the uncured state.
Shelf life	Length of time a material can be stored without losing specific properties
Spheroidal cast iron	See **Ductile cast iron**
Steel	A malleable, ductile alloy of iron and carbon, having a carbon content varying between close to 0 per cent and 1.8 per cent (above which the alloy is known as cast iron)
Substrate	A material or structure to which an adhesive is applied for a purpose such as bonding
Tapered plate	A gradual reduction in the thickness of an FRP plate, achieved by reducing the number of plies of fibre
Toughness	A measure of the ability of a material to absorb energy, defined as the work required to rupture a unit volume of material
Unidirectional	A fibre arrangement in which all the fibres are aligned in a single direction
Vacuum bag consolidation	A process for moulding laminates in which a sheet of flexible material is placed over the lay-up on the mould and sealed. A vacuum is applied between the sheet and lay-up. The entrapped air is pulled out of the lay-up and removed by the vacuum. Atmospheric pressure provides the consolidation pressure
Vacuum infusion	A composite manufacturing method in which resin is drawn into a lay-up of dry fibres under vacuum. Can be used *in situ* to strengthen an existing structure
Wet lay-up	A method of manufacturing a reinforced-fibre laminate in which layers of reinforcement are impregnated by applying a liquid resin system as they are laid up
Wrought iron	A malleable and almost pure form of iron (less than 0.1 per cent carbon) incorporating threads of slag. Produced either by direct reduction of iron ore in the presence of carbon or by heating cast iron in the presence of oxygen. When worked, the iron gains strength. This form of iron is no longer in production
Yarn	Continuously twisted fibres or strands that are suitable for weaving into fabrics

Nomenclature

Subscripts

0	before strengthening	i	first installation of the metallic structure
1	after strengthening	matrix	matrix material in the FRP
a	adhesive	R	released state if the adhesive bond was not present
c	compressive limit	s	metallic substrate
f	FRP strengthening material	t	tensile limit
fibre	fibre material in the FRP	u	ultimate state

General notation

$\overline{X}$ an overbar indicates a limiting value of a variable

Roman characters

a	crack length in the substrate	E_{eff}	effective modulus of elasticity, allowing for creep
a_1, a_2, a_3	flexibility parameters in adhesive joint analysis	E_f	modulus of elasticity of FRP strengthening material
$\mathbf{A_1}$	membrane biaxial stiffness matrix, for the strengthened section	E_{fibre}	modulus of elasticity of fibres in FRP
a_d	design values for geometric data	E_i	modulus of inner adherends (for double lap-shear test)
A_f	cross-sectional area of FRP strengthening material	E_S	modulus of elasticity of metallic substrate
$\mathbf{A_f}$	membrane biaxial stiffness matrix, for the FRP material alone	E_{matrix}	modulus of elasticity of matrix in FRP
A_s	cross-sectional area of metallic section	E_{xf}	Young's Modulus of FRP in x-direction
$\mathbf{A_s}$	membrane biaxial stiffness matrix, for the substrate alone	E_{yf}	Young's Modulus of FRP in y-direction
b	breadth	f_1, f_2	flexibility parameters in adhesive joint analysis
$\mathbf{B_1}$	flexural biaxial stiffness matrix, for the strengthened section	G	shear modulus
		G^*	equivalent shear modulus in biaxial analysis
b_a	breadth of adhesive	G_a	shear modulus of adhesive
b_f	breadth of FRP plate	G_f	shear modulus of FRP
c	anchorage length of FRP beyond theoretical curtailment point	G_1	fracture energy release rate
$C_1, C_2, C_3,$	boundary condition constants in the adhesive joint analysis	$\overline{G_{IC}}$	critical fracture energy release rate
		G_S	shear modulus of substrate
C_4, C_5	simplifying functions in the adhesive joint analysis	I_1	second moment of area of the composite, strengthened section
c_1	anchorage length required for stress development in substrate	I_f	second moment of area of FRP strengthening
c_2	anchorage length required for variability	I_s	second moment of area of metallic section
		I_{ww}	warping constant for strengthened section
c_3	anchorage length required for delamination damage	I_{ys}	second moment of area of metallic section about the minor axis
c_a	parameter in lap-shear specimen calculation	I_{yf}	second moment of area of FRP about the minor axis of the strengthened section
$\mathbf{D_1}$	coupling biaxial stiffness matrix, for the strengthened section	I_{yy}	second moment of area of strengthened section about the minor axis
$D_{11}, D_{12}, D_{22}, D_{66}$	flexural rigidities of a strengthened plate	j	Findley's law coefficient, dependent on permanent stress
E^*	equivalent axial Young's Modulus in biaxial analysis	J	torsion constant for strengthened section
E_0	time-independent modulus in creep calculations	k	Findley's law coefficient, a function of environmental conditions
E_o	modulus of outer adherends (for double lap-shear test)	k	parameter in lap-shear specimen calculation
E_a	modulus of elasticity of adhesive	k'	parameter in lap-shear specimen calculation
E_a'	parameter in lap-shear calculation		

ℓ	length of circular transition due to curvature of plate
L	buckling length of member
L_a	overlap length in lap-shear specimen
m	moisture content of adhesive
M	bending moment applied to the strengthened section
$\mathbf{M}$	stress resultant vector (bending moments per unit width) for biaxial loading
$\overline{M}$	limiting bending moment applied to the strengthened section
M_0	initial bending moment due to permanent loads in unstrengthened member
M_{cr}	buckling capacity of strengthened member
M_f	bending moment in the FRP strengthening material, about the centroid of the strengthening
M_f^*	bending moment in the FRP strengthening material, about the interface between the strengthening and the adhesive
M_{f0}	initial bending moment in the FRP
M_{PS}^*	particular solution moment in adhesive analysis
M_S	bending moment resulting from loads applied to the strengthened member after the adhesive set, assuming only the metallic member carries the load
M_{su}	ultimate moment capacity of the unstrengthened section (substrate alone)
M_t	bending moment due to restrained thermal actions at the operating temperature T.
M_{t0}	bending moment due to restrained thermal actions at temperature T_0.
M_u	ultimate moment capacity of the strengthened section
n	empirical exponent in the hygrothermal equation
N	axial force applied to the strengthened section
$\mathbf{N}$	stress resultant vector (forces per unit width) for biaxial loading
$\overline{N}$	limiting axial force applied to the strengthened section
N_0	Initial axial force due to permanent loads in unstrengthened member
n_f	number of fatigue cycles to failure
N_f	axial force in the FRP strengthening material
N_{f0}	initial force in the FRP at the time of strengthening, due to prestress.
$N_f\big|_{x=0}$	axial force in FRP at $x=0$ (the position of a strain discontinuity).
N_{PS}	particular solution axial force in adhesive analysis
N_s	axial force resulting from loads applied to the strengthened member after the adhesive set, assuming only the metallic member carries the load
N_t	axial force due to restrained thermal actions at the operating temperature T.
N_{t0}	axial force due to restrained thermal actions at temperature T_0.
$N_{x,cr}$	local buckling strength of a simply-supported strengthened plate
P	load applied to lap-shear test specimen
R	radius of curvature of FRP strengthening material
$\mathbf{R}$	generic description of resistance of structure
s	separation between FRP plates
$\mathbf{S}$	generic description of destabilising loads
t	time (for creep calculations)
T	operating temperature
T_0	temperature at time of installing strengthening
t_a	thickness of adhesive
T_A	minimum allowable difference between the glass transition temperature and the t_f thickness of FRP
t_f	thickness of FRP
T_f	temperature of FRP
T_{f0}	temperature of FRP at time of installing strengthening
T_g	glass transition temperature
ΔT_g	shift in glass transition temperature for 100 per cent saturation of the FRP
t_i	thickness of inner adherends (for double lap-shear test)
T_i	temperature at time of installing metallic member
t_o	thickness of outer adherends (for double lap-shear test)
T_{ref}	reference temperature at which a material property is known
t_S	thickness of metallic substrate
T_S	temperature of metallic substrate
T_{S0}	temperature of metallic substrate at time of installing strengthening
u	strain energy per unit length of a general section
U	total strain energy in a beam
u_1	strain energy per unit length of the strengthened section
u_2	strain energy per unit length of the unstrengthened section
u_3	parameter in lap-shear specimen calculation
u_4	parameter in lap-shear specimen calculation
V	applied shear force
V_f	shear force in the FRP strengthening material
V_{fibre}	fibre volume fraction
V_s	shear force resulting from loads applied to the metallic member after the adhesive set
w_D	applied dead load (in worked example)
w_L	applied live load (in worked example)
x	position along the beam
X	generic material property (in the hygrothermal equation)
$x_{L,i}$	generic load variables
$x_{M,i}$	generic material variables
y	vertical position within section
y_f	distance from the centroid of the FRP to the adhesive interface
y_{g0}	distance of centroid of unstrengthened section from the tension face of a beam
y_{g1}	distance of centroid of strengthened section from the tension face of a beam
y_s	distance from the centroid of the beam to the adhesive interface
z	internal lever arm between the centroid of the unstrengthened section and the centroid of the FRP

Greek characters

α_f	coefficient of thermal expansion for FRP
α_{fibre}	coefficient of thermal expansion of fibres in FRP
$\alpha_{longitudinal}$	coefficient of thermal expansion of FRP in the longitudinal direction
α_{matrix}	coefficient of thermal expansion of matrix resin in FRP
α_s	coefficient of thermal expansion for metallic substrate
$\alpha_{transverse}$	coefficient of thermal expansion for the transverse direction
β	flexibility parameter in adhesive joint analysis
β_s	parameter in single lap-shear analysis
γ_{f3}	partial factor to allow for uncertainty in design
γ_{fL}	load partial factor
$\gamma_{fL,i}$	load partial factor on the ith applied load
γ_m	material partial factor
γ_{me}	material partial factor for environmental effects
γ_{mf}	material partial factor describing variability of the constituent materials used
$\gamma_{M,i}$	material partial factor on the ith material
γ_{mm}	material partial factor describing variability due to manufacturing process by which an FRP is produced
γ_{mt}	material partial factor for time related effects
γ_{mv}	material partial factor for variability
γ_{xy}	shear strain in FRP
$\bar{\gamma}_{xy}$	limiting shear strain in FRP
$\boldsymbol{\varepsilon}$	strain vector for biaxial loading
ε_0	time-independent strain (for creep calculations)
ε_1	strain at the centroid of the strengthened section
ε_f	strain in FRP
ε_{f0}	the initial difference in strain between the FRP and the substrate material, defined at the time of strengthening
$\Delta\varepsilon_f$	increment of strain in FRP
$\bar{\varepsilon}_{fc}$	limiting compressive strain for FRP
$\bar{\varepsilon}_{ft}$	limiting tensile strain for FRP
ε_{fp}	prestrain in FRP
ε_{fR}	strain in the FRP if bond across the adhesive joint is released
$\Delta\varepsilon_{fs}$	lack-of-fit strain between the FRP and the substrate if there was no adhesive bond
ε_s	strain in the metallic substrate
$\Delta\varepsilon_s$	increment of strain in the metallic substrate
ε_{s0}	strain in the metallic substrate, before strengthening
$\bar{\varepsilon}_{sc}$	limiting compressive strain for metallic substrate
ε_{sR}	strain in the substrate if bond across the adhesive joint is released
$\bar{\varepsilon}_{st}$	limiting tensile strain for metallic substrate
ε_u	ultimate strain
ε_x	FRP strain in x-direction
$\bar{\varepsilon}_x$	limiting FRP strain in x-direction
ε_y	FRP strain in y-direction
$\bar{\varepsilon}_y$	limiting FRP strain in y-direction
ε_{yg0}	strain in the tension face of the metallic section at the time of strengthening.
χ	parameter in lap-shear calculation
Ω	parameter in lap-shear calculation
$\boldsymbol{\psi}$	curvature vector for biaxial loading
ψ_1	curvature in strengthened section
ψ_{fR}	curvature in the FRP if bond across the adhesive joint is released
$\Delta\psi_{fs}$	lack of fit curvature between the FRP and the substrate if the adhesive bond was not present.
ψ_{sR}	curvature in the substrate if bond across the adhesive joint is released
λ	relative flexibility parameter in adhesive joint analysis
λ_d	parameter in double lap-shear specimen calculation
λ_s	parameter in single lap-shear specimen calculation
ν_a	Poisson's ratio of adhesive
ν_f	Poisson's ratio of adherend in lap-shear test
ν_{fibre}	Poisson's ratio of fibres in FRP
ν_{matrix}	Poisson's ratio of matrix in FRP
ν_s	Poisson's ratio of metallic substrate
ν_{xyf}	Poisson's ratio of FRP in x-direction
ν_{yxf}	Poisson's ratio of FRP in y-direction
$\Delta\theta$	angle turned through by FRP
σ	peel stress in the adhesive, normal to the interface plane
$\bar{\sigma}$	characteristic strength of adhesive
σ_1	maximum principal stress in adhesive
σ_f	stress in the FRP
σ_{fp}	prestress in FRP
σ_{max}	maximum peel stress in the adhesive
σ_n	maximum stress in a fatigue cycle
$\bar{\sigma}_n$	static failure stress corresponding to σ_n
σ_S	stress in the metallic substrate
σ_{S0}	stress in the metallic substrate, before strengthening
σ_t	permanent stress resulting in creep
σ_{tu}	ultimate tensile strength of substrate
σ_{fx}	FRP stress in x-direction
$\bar{\sigma}_{fx}$	strength of FRP in x-direction
σ_{fy}	FRP stress in y-direction
σ_y	yield strength of ductile metallic substrate
$\bar{\sigma}_{fy}$	strength of FRP in y-direction
τ	shear stress in the adhesive, parallel to the interface plane
τ_{max}	maximum shear stress in the adhesive joint
τ_{xy}	shear stress in FRP
$\bar{\tau}_{xy}$	shear strength of FRP

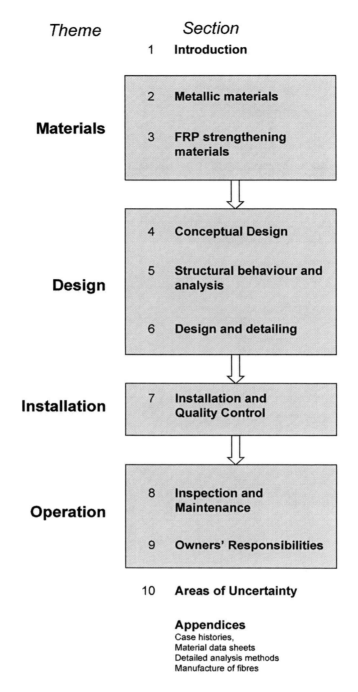

Figure 1.1 *Report structure*

How to use the report

The structure of this report mirrors the four principal stages in an FRP strengthening project, as illustrated in Figure 1.1 opposite.

The introduction (Chapter 1) presents the scope of the document, and introduces the materials and techniques addressed in it.

Chapters 2 and 3 provide technical information on the *materials* that are encountered when designing FRP strengthening schemes for metallic structures. They are introductory and contain material that may already be familiar to some readers. Chapter 2 describes metallic structural materials (cast iron, wrought iron and steel) and associated typical forms of construction, while Chapter 3 describes *FRP strengthening systems*, including the FRP strengthening material, the adhesive bond, and manufacturing methods of the strengthening component.

Design guidance is provided in Chapters 4 to 6. Chapter 4 covers conceptual design, including typical forms of externally bonded FRP strengthening and criteria for selecting a strengthening system. Detailed design is split into analysis methods (Chapter 5) and the design framework (Chapter 6), including factors of safety and detailing guidelines. Chapter 5 includes two step-by-step procedures for the guidance of designers.

Installation procedures for FRP strengthening on a metallic structure are described in Chapter 7.

Chapters 8 and 9 address *management and operation aspects* of an FRP strengthened metallic structure. Chapter 8 outlines a typical inspection and maintenance regime. Chapter 9 covers various issues that are of interest to the owners of structures.

Additional information is provided at the end of the report. Chapter 10 lists current areas of uncertainty, to guide future research on the subject. Case histories are included in the appendices, together with further information on FRP materials and detailed analysis methods.

1 Introduction

Fibre-reinforced polymer (FRP) strengthening is a powerful technique for extending the life of structures. FRP strengthening allows existing metallic structures to be upgraded when they no longer possess adequate structural strength or stiffness due to corrosion, fatigue, damage or a change of use or requirements.

Externally bonded FRP strengthening is particularly attractive where there are severe access constraints or high costs associated with installation time. This is usually the case when the structure to be strengthened is in continuous service and the interruption of its availability to permit the installation of strengthening entails heavy disruption costs.

The capacity of FRP strengthening to extend the life of historic structures with minimum disruption to users makes for genuinely sustainable engineering solutions.

1.1 AIMS OF THIS REPORT

Externally bonded FRP strengthening for metallic structures is a rapidly developing technique, but at the time of writing very little guidance is available for its use. CIRIA commissioned this report to describe the current best practice for strengthening metallic structures using FRP. It provides information for a wide audience, including:

- clients and the owners of structures
- contractors responsible for installation
- consultants who design strengthening solutions
- manufacturers and suppliers of FRP strengthening systems.

The principal aims of the report are to:

- promote the use of externally bonded FRP strengthening for metallic structures
- describe the appropriate use of the technique
- provide guidance for structural design
- provide guidance for the correct implementation of strengthening schemes
- provide guidance on the future maintenance of strengthened structures.

It must be recognised that the use of externally bonded FRP for strengthening metallic structures is a developing technology. This is a best practice report, which brings together the best available knowledge on the subject, rather than present the results of new research of a fundamental nature. It thus identifies aspects for which definitive guidance cannot at present be offered and it does not attempt to fill in this knowledge.

CIRIA commissioned FaberMaunsell and the University of Surrey to act as research contractors. CIRIA appointed a steering group, comprising industry experts, to advise on the content and technical sufficiency of the report. The work has also benefited from an in dustry-wide, international consultation exercise and has been reviewed by leading experts in the field.

1.2 WHAT ARE FIBRE-REINFORCED POLYMERS?

Fibre-reinforced polymer (FRP) materials combine high-strength, high-modulus fibres with a low-modulus polymeric matrix material. The result is a high-strength, high-stiffness composite in which the matrix material ensures load transfer between the fibres. Fibres can be of glass, aramid or carbon. A wide range of FRP composites can be produced by choosing appropriate fibres and matrix materials, and by selecting the arrangement of fibres within the composite. In this way the properties of FRP materials can be tailored to suit a particular application.

Fibre-reinforced polymer (FRP) materials are widely used in the aerospace and automotive industries. They have also been used in the construction industry for 30 years, as described in CIRIA publication C564 *Fibre-reinforced polymer composites in construction* (Cripps, 2001). Construction applications include:

- all composite structures (such as bridges, modular buildings and cladding systems)
- composites used in conjunction with conventional civil engineering materials, such as:
    - load-bearing FRP panels
    - reinforcement for reinforced concrete
    - replacing of supplementing the tensile component of a reinforced concrete beam
    - wrapping reinforcement, to improve the buckling and compressive capacity of reinforced concrete columns
    - externally bonded strengthening for reinforced concrete or metallic beams.

1.3 WHAT IS EXTERNALLY BONDED FRP STRENGTHENING?

FRP composites are particularly suited to the strengthening and repair of existing metallic structures, since they can be bonded to an exposed surface of the structure. There are three distinct elements to be considered in a structure strengthened using FRP:

- the original metallic structure
- the FRP composite strengthening material
- the structural adhesive bonding the metallic substrate and the FRP strengthening material.

The initial development of FRP strengthening techniques largely focused on concrete structures. The first externally bonded FRP strengthening application was the Ibach Bridge in Switzerland in 1991. The first application of FRP strengthening to metallic structures took place five years later, in 1996.

"Strengthening" is used in this report as a generic term to describe all applications of externally bonded strengthening, including stiffening or repairing a structure, extending its fatigue life, or increasing its load capacity.

1.4 WHAT METALLIC STRUCTURES CAN BE STRENGTHENED USING FRP STRENGTHENING?

Metallic structures include bridges, metal-framed buildings, covered ways, pipes and vessels. They exhibit a variety of structural forms that reflect the development of metallic structural materials from cast iron, through wrought iron, to steel. Any of these structures could potentially require strengthening.

The reasons for strengthening a metallic structure include:

- increasing the load capacity of a structure to cater for loading beyond that envisaged in the original design
- replacing material lost due to corrosion, material impaired by impact, or material property degradation
- making good construction deficiencies, such as design errors, sub-standard materials, and poor workmanship
- stiffening the structure to reduce deflection and increase buckling capacity
- extending the fatigue life of a structure.

1.5 AVAILABLE DESIGN GUIDANCE DOCUMENTS

Design guidance and codes are well established for the design of metallic structures, and their assessment. At the time of writing, however, there are no mature design codes for externally bonded FRP strengthening, applicable to either metallic or concrete structures.

The following guidance documents are relevant to the report.

Composites

- *Structural design of polymer composites. Eurocomp design code and handbook* (Clarke, 1996)
- BS 4994:1987 *Specification for design and construction of vessels and tanks in reinforced plastics*
- Offshore Standard DNV-OS-C501 *Composite components* (DNV, 2003)
- CIRIA C564 *Fibre-reinforced polymer composites in construction* (Cripps, 2001)

Adhesive joints

- *A guide to the structural use of adhesives* (IStructE, 1999)
- BA 30/94 *Strengthening of concrete highway structures using externally bonded plates* (Department of Transport, 1994)

Externally bonded FRP, applied to concrete

- TR55 *Design guidance for strengthening concrete structures using fibre composite materials* (Concrete Society, 2000)
- TR57 *Strengthening concrete structures with fibre composite materials: acceptance, inspection and monitoring* (Concrete Society, 2003)
- fib bulletin 14 *Externally bonded FRP reinforcement for RC structures* (fib, 2001)
- ACI 440.2R-02 *Guide for the design and construction of externally bonded FRP systems for strengthening concrete structures* (ACI, 2002)

Externally bonded FRP, applied to metallic structures

- ICE design and practice guide *FRP composites – life extension and strengthening of metallic structures* (Moy, 2001b)

This report complements Concrete Society TR55 (Concrete Society, 2000) and extends the scope of the ICE design and practice guide (Moy, 2001b). For background on the general use of FRP composites in construction see CIRIA C564 (Cripps, 2001).

1.6 BIBLIOGRAPHY

ACI (2002). *Guide for the design and construction of externally bonded FRP systems for strengthening concrete structures*. ACI 440.2R-02, American Concrete Institute, Detroit

Clarke, J L, ed (1996). *Structural design of polymer composites. Eurocomp design code and handbook*. E & FN Spon, London

Concrete Society (2000). *Design guidance for strengthening concrete structures using fibre composite materials*. Technical Report 55, Concrete Society, Crowthorne

Concrete Society (2003). *Strengthening concrete structures with fibre composite materials: acceptance, inspection and monitoring*. Technical Report 57, Concrete Society, Crowthorne

Cripps, A, ed (2001). *Fibre-reinforced polymer composites in construction*. C564, CIRIA, London

Department of Transport (1994). *Strengthening of concrete highway structures using externally bonded plates*. BA 30/94 (DMRB vol 3, sec 3, pt 1), HMSO, London

DNV (2003) *Composite components*. Offshore Standard DNV-OS-C501, Det Norske Veritas, Oslo

fib (2001). *Externally bonded FRP reinforcement for RC structures*. fib bulletin 14, Fédération internationale du béton, Lausanne

IStructE (1999). *A guide to the structural use of adhesives*. SETO, London

Moy, S S J, ed (2001b). *FRP composites – life extension and strengthening of metallic structures*. ICE design and practice guide, Thomas Telford, London

British Standard

BS 4994:1987. *Specification for design and construction of vessels and tanks in reinforced plastics*

2   Metallic materials

> A wide range of metallic materials has been used in construction and, as new materials became available, structural form also developed. Any of these materials and structures may require strengthening. An understanding of the metallic structure is important before strengthening works be planned since each material has its own characteristics that determine how externally bonded FRP should be used.
>
> This chapter presents an overview of metallic structural materials. It describes the most commonly used structural metals (cast iron, wrought iron and steel) and gives a brief history of their development. The characteristics of these metals are presented including their microstructure, typical defects and mechanical properties. Guidance on the identification and characterisation of metallic materials and on their structural assessment is provided.
>
> A more detailed description of structural metallic materials can be found in Bussell (1997) and in the references given at the end of the chapter.

2.1   AN HISTORIC OVERVIEW OF METALLIC STRUCTURES

Before planning a strengthening scheme, the nature of the metallic structure to be strengthened must be determined. Of particular importance are the materials and fabrication methods available at the time of construction. The following sections chart the development of metallic structures from cast iron, through wrought iron to steel structures. This is summarised in Figure 2.1.

2.1.1   Cast iron

The use of structural cast iron developed at the end of the 18th century. Cast iron columns were first used in churches in the 1770s, but were used more widely in textile mills from the 1790s onwards. The first cast iron bridge, Ironbridge, was built in 1779. The first iron-framed building, Ditherington Flax Mill, Shrewsbury, dates back to 1796. Cast iron was subsequently extensively used in the construction of bridges and large buildings, such as major public buildings in London and textile mills in northern England.

The forms of construction during this period were governed by the properties of the cast iron available. For spans less than 15 m, cast iron primary girders were combined with transverse secondary girders or a timber deck. For spans up to 35 m, cast iron beams were bolted together and trussed with wrought iron rods. Cast iron arch bridges of up to 60 m span are still in use (Figure 2.2). Jack arch construction was widely used to form the floors of industrial buildings such as textile mills and warehouses (Figure 2.3), to form road decks and in cut and cover construction (used extensively for the sub-surface London Underground lines). Jack arch construction (in which cast iron was later displaced by wrought iron and eventually steel) remained popular until around 1870.

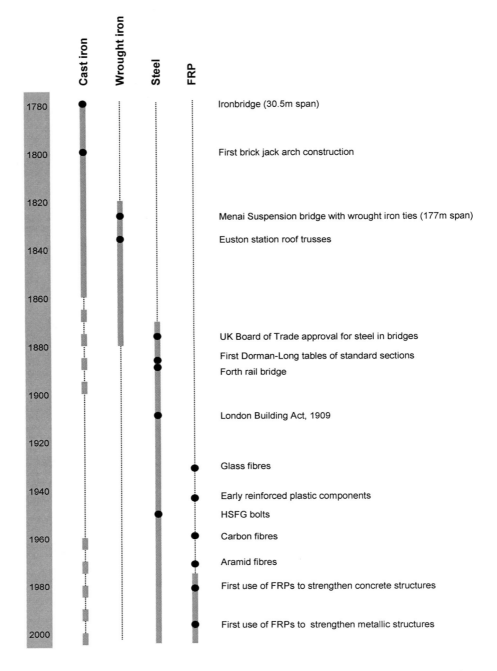

Figure 2.1 *The historic development of metallic and FRP materials for structural engineering*

Cast iron structures were generally connected by direct bearing or mechanical interlock. Connections inspired by timber or masonry connections such as spigot and socket and wedge connections were commonly used. Cast iron beams were frequently seated on capitals at the head of cast iron columns.

Following the collapse of the cast-iron Dee Bridge in 1847, wrought iron replaced cast iron for railway bridge construction. Nevertheless, cast iron still finds use in structures. Modern grey cast iron is similar to historic cast iron, but is made to higher standards.

Ductile cast iron (also known as spheroidal or nodular graphite cast iron) is a modern material with a different microstructure to grey cast iron. Whereas the latter is brittle, ductile cast iron has similar properties to carbon steel. Ductile cast iron is used for structures where the formability or thickness of cast iron is essential, such as tunnel linings, pipes, manhole covers, thick pipes and space frame nodes. It is also used to replace existing grey cast iron elements, and in modern building construction.

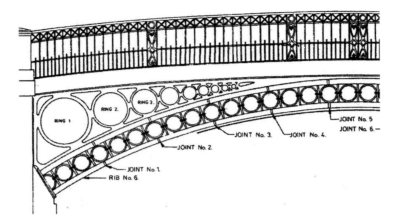

Figure 2.2   *Typical elevation of a cast iron arch bridge*

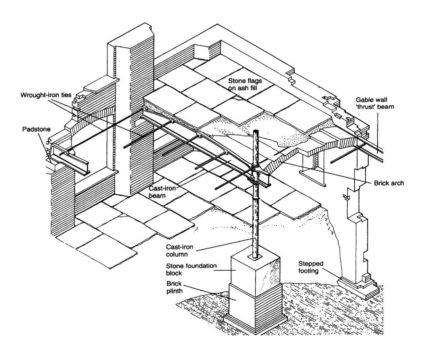

Figure 2.3   *Typical textile mill construction (from Swailes and Marsh, 1998)*

2.1.2   Wrought iron

Wrought iron took over from cast iron as a higher performance structural material between the mid-1840s and early 1850s. The superior tensile strength of wrought iron allowed it to be used in truss and girder bridges, for example Brunel's Chepstow Bridge (built in 1852 and dismantled in 1962). It was also widely used in roof trusses, driven by the requirement for long spans during the first half of the 19th century, Euston station roof (1837, demolished 1962) being a good example.

Wrought iron was frequently used in combination with cast iron. For example, in textile mills wrought iron roof trusses were combined with cast iron columns and jack arch floor beams, and wrought iron tie rods were used in jack arch construction. Wrought iron beams replaced cast iron in brick jack arch structures.

Wrought iron structures were generally connected by riveting or, less frequently, by bolting. Shrink-ring connections were also used to hold members together. A large proportion of the UK's railway bridges of the time were constructed from wrought iron. Beam sections were built-up from wrought iron plates and bars, with rivets to join them (Figure 2.4).

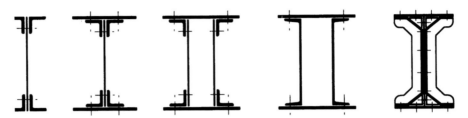

Figure 2.4   *Built-up wrought iron beam sections (from Bussell, 1997)*

Wrought iron was rapidly superseded by mild steel at the end of the 19th century. Production abruptly ceased, and wrought iron is no longer available in the form produced in the 19th century.

2.1.3   Early steel

Commercial steel was being produced in Britain by Bessemer's steel-making process by about 1860, but it is rare to find steel used in structures completed before 1890. For structures built after 1900, however, it is unusual to find any material other than steel.

The development of steel construction was greatly aided by the adoption of standard sections and design codes. The first tables of standard sections were produced by Dorman-Long in 1887, and the first British Standard (BS 1) for steel sections was published in 1901, followed by the London Building Act in 1909.

Early steel construction was fabricated from sheet and plate material connected by rivets or black bolts. The first steel skyscrapers were constructed in the USA in the early 1880s, followed by the first British steel-framed building in 1896. The first major steel structure was the Forth Railway Bridge in 1889.

2.1.4   Modern steel

Two developments in the 1950s mark the era of "modern" steel. Due to the continuous improvement in the quality of available steel it became possible to replace riveting by welding in the fabrication of structural sections, although on-site welding of connections was not used with confidence until the 1970s. Secondly, high-strength friction-grip bolts were introduced, replacing black bolts, and this simplified the formation of connections.

Steel-concrete composite construction has become common in highway bridge construction, where plate or box girders act compositely with a reinforced or prestressed concrete deck slab, and in building floors, in which a concrete slab acts compositely with profiled steel sheeting, which doubles as permanent formwork. Concrete-filled or conrete-encased beams and columns constitute another application of composite construction, often motivated by the need to increase fire resistance or load carrying capacity within a compact envelope.

2.1.5 Aluminium and stainless steel

Aluminium and stainless steel have both found recent structural use. The structural development of aluminium was largely driven by its use in the aerospace industry, starting in the 1920s, and its structural use in construction was promoted by an IStructE report published in 1960.

Structural aluminium and stainless steel are both potential candidates for strengthening. The principles of the techniques described for steel are equally applicable to aluminium and stainless steel, and therefore their application to these materials will not be discussed further in this report. IStructE (1996) contains details on stainless steels and aluminium alloy properties.

2.1.6 Metallic structures requiring strengthening

The development of metallic structures has left a wide range of metallic materials and forms in existing structures, from historic 19th-century structures to modern steel-framed construction. Many of these structures carry loads far greater than those for which they were originally intended and are able to do so only because of the conservative design criteria of the time. Increased loading and/or fatigue requirements may also be demanded of an existing structure, either as a result of a change of use or due to modernisation programmes for the networks that the structure serves.

Railway bridges

The UK mainline railway system includes a large number of bridges, of which around 40 per cent are metallic and 25 per cent are carried over railways. Many of the UK's railway bridges are of wrought iron, comprised of fabricated plate girders connected by rivets. The bridges commonly consist of a grillage of main longitudinal girders, cross girders and infill girders, carrying either a timber or steel deck, on which is laid the stone ballast and the track. Heavier live loads and regular train axle spacing mean that railway bridges tend to be more susceptible to vibration and fatigue loading than their highway counterparts. Bridges more than 50 years old are generally riveted and may be badly corroded on all surfaces. In particular, bottom flanges may be badly pitted and rusted and rivets may be loose.

The strengthening of bridges can cause considerable disruption to railway services, so FRP strengthening techniques that can be applied in a minimum amount of time are of significant benefit.

Underground railways

Many of the cut-and-cover sections of the London Underground system are roofed by jack arch construction, comprising cast iron beams. Elsewhere, cast iron struts are used to prop retaining walls and ventilation shafts. Bored tunnel sections make extensive use of cast iron tunnel linings. Above-ground development, including new buildings and increased traffic loads, place increasing demands upon the underground infrastructure.

Highway bridges

In the early days of the road network steel tended to be used for long-span structures of riveted truss and plate girder construction. The development of welded steel girder construction led to the use of composite I-beams for short to medium spans and box girders for long spans and moveable bridges. In the UK, 10–15 per cent of bridges are believed to be metallic. Metallic highway bridges tend to be younger than railway bridges, but the use of de-icing salts can result in faster corrosion, particularly if there have been lapses in the maintenance regime. A similar situation exists elsewhere. For example, in 2000 the US Federal Highway Administration had about 52 000 structurally deficient steel bridges.

Growing volumes of traffic and increasing vehicle weights have led to a continual need for road and bridge strengthening, sometimes accompanied by widening. This has affected many comparatively young steel structures such as the first Severn Bridge, Avonmouth Bridge, Tamar Bridge, Tees Viaduct and Midland Links viaducts.

The strengthening of highway structures can seriously disrupt traffic, so techniques such as FRP that can be applied in less time are beneficial.

Waterways

Waterways include bridges, locks, aqueducts and tunnels. Although many of them are in masonry, the canal network also includes cast iron bridges of relatively short span, many of which are of historical significance. Around 20 per cent of British Waterways' structures are of metallic or metal-concrete composite construction.

Building structures

Before 1900 a very large number of cast-iron-framed, brick jack-arch structures were built as public buildings, textile mills and warehouses. Many of these still stand, and careful refurbishment allows these structures to be adapted for alternative modern uses.

Steel-framed construction replaced cast iron early in the 20th century, and is now widespread, for example in office buildings, car parks, station canopies, power stations and factories. The use of these structures will often need to change, necessitating structural modification, such as the introduction of openings in floors for the insertion of services and lift shafts. Older building frames are predominantly of bolted or riveted construction, whilst more recent structures make greater use of shop-welded frame components in conjunction with bolted site connections.

Industrial structures

A large number of steel structures, such as processing plants, tanks, marine piers and jetties, are constructed from metallic materials. One sector where FRP strengthening is already used is the offshore oil and gas production industry. FRP strengthening material is also applied to ships.

2.2 THE STRUCTURAL CHARACTERISTICS OF METALLIC MATERIALS

Each of the metallic materials that can be strengthened using externally bonded FRP has distinctive material properties. It is essential to understand the characteristics of the material being strengthened before undertaking the design of a strengthening scheme. Table 2.1 gives indicative properties for cast iron, wrought iron and steel.

Table 2.1 *Typical properties of structural metallic materials*

		Cast iron		Wrought iron	Steel	
		Historic	Modern (BSI, 1997a)		Historic (pre-1950)	Modern (post-1950)
Modulus (GPa)	tensile	66–94*	100–145	154–220	200–205	200–210
	compressive	84–91*				
Strength (MPa)	tensile	65–280	150–400	278–593**	286–494	275–355
	compressive	587–772	600–1200	247–309		
Elastic limit (MPa)		See Figure 2.5		154–408	278–309	275–355
Ultimate strain (%)		Figure 2.5	0.5–0.75 (tension)	7–21	18–20	18–25
Poisson's ratio		0.25	0.26	0.25	0.26–0.34	0.30
Density (kg/m³)		7050–7300		7700	7840	
Coefficient of thermal expansion ($10^{-6}/^{0}$C)		10–11		12	12	

* Secant modulus, due to non-linear σ-ε curve.

** UTS is parallel to the grain. UTS perpendicular to the grain is about two-thirds to three-quarters of this value

All values in this table are representative. Reference should be made to Bussell (1997) and Highways Agency (2001) for further information, together with any test data.

2.2.1 Grey cast iron

Grey cast iron is an alloy of iron containing 2.5–4 per cent by weight of carbon (graphite) and elements such as silicon, phosphorous, manganese and sulphur, all of which significantly influence its material properties. After casting, grey cast iron is cooled slowly, allowing the graphite to form discrete flakes, which act as stress raisers. Cast iron is weak in tension, brittle, fatigue-sensitive and exhibits a significant size effect.

Defects

In addition to variation in the internal microstructure, grey cast iron is characterised by macrostructure defects produced during casting (Swailes and Marsh, 1998):

- blow-holes, due to poor venting of the mould, of up to 10 mm in diameter
- weaker, coarser material in the centre and residual stresses, due to differential cooling of the section
- contamination by sand that became detached from the mould
- cold joints, due to interruptions in casting
- cold-spots, where an earlier splash of molten iron has cooled and solidified without being absorbed by further molten iron
- large variations in sectional thickness.

Material properties

Cast iron has a non-linear stress-strain response, due to the graphite inclusions. A typical stress-strain curve for a grey cast iron is shown in Figure 2.5.

Cast iron's tensile stress-strain curve has no distinct point separating the elastic and elasto-plastic portions, the curve forming a seamless whole. This applies even in the case of ductile cast iron. The term "yield point" in its conventional sense is therefore inapplicable to cast irons. Ultimate failure of the cast iron is approximated by a maximum shear strain energy (Von Mises) criterion in compression and a principal tensile stress (Rankine) criterion in tension.

Where the non-linear response of the cast iron is important, a curve can be fitted to the stress-strain response. For example, the following relationship was used to characterise tests on cast iron members at Sheffield University (Moy, 1999):

$$\sigma = E\left(\varepsilon - a\,\varepsilon^{b}\right) \tag{2.1}$$

where $a = 5.5$ and $b = 1.4$. With $E = 180$ MPa, failure occurred at a stress of 180 MPa, and a strain of 0.7 per cent.

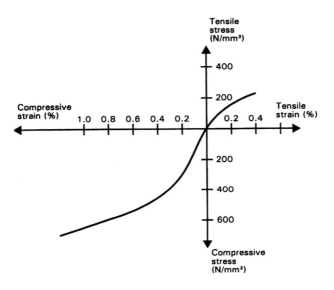

Figure 2.5   *A typical stress-strain response for grey cast iron (from Bussell, 1997)*

Cast iron components are susceptible to fatigue degradation due to casting flaws and high residual shrinkage stresses. Angus (1976) gives a detailed discussion of the impact properties and fatigue resistance of cast iron.

The heavy mass and relatively low design stresses used for cast iron members mean they generally perform well in fire. Under severe loading conditions, however, thermal movements can overload cast iron members, leading to failure. Under sustained high temperatures, cast iron columns will eventually buckle. A further concern is that cast iron members may fail due to thermal shock loads imposed by fire-fighting equipment.

Cast iron has good corrosion resistance, as silica in the moulding sand coats the surface of the casting. Cut or fractured surfaces, however, rust rapidly.

Design values for the material properties of historic cast iron

Cast iron passed out of use before design codes became widespread (Figure 2.1). For design purposes, cast iron is usually assumed to be linear-elastic. This is reasonable, as the material is constrained to low working stress by a large safety factor. Elastic bending theory is applied using a suitable failure criterion. A large factor of safety is applied to historic cast iron, to allow for the possibility of defects, the brittle nature of the material and its sensitivity to fatigue loading.

BD 21/01 (Highways Agency, 2001a) suggests permissible stress values for the assessment of cast iron highway structures. These stresses are represented by the stress envelopes shown in Figure 2.6, which plot the permissible dead load stress against the permissible live load stress. They include a weighting to increase the effect of dynamic forces, and the possibility of fatigue failure. Although fatigue failure is not so relevant in buildings, no better data is available at the present time. Where more specific values are required, tests should be carried out on samples from the cast iron.

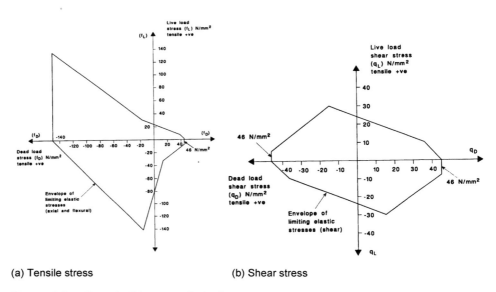

(a) Tensile stress   (b) Shear stress

Figure 2.6 *Permissible stress limits for cast iron in BD 21/01 (Highways Agency, 2001a; reproduced from Bussell, 1997)*

Design values for the material properties of modern cast iron

Improvements to the manufacturing processes mean that modern grey cast iron has a higher tensile strength than its historic equivalent. BS EN 1561:1997) specifies different grades of modern grey cast iron and the material properties associated with these different grades.

2.2.2   **Ductile cast iron**

Ductile cast iron contains spherical graphite nodules, hence its alternative names of spheroidal or nodular cast iron. Under tensile load, the spherical inclusions do not act as stress raisers to the same extent as the graphite flakes in grey cast iron. Consequently, spheroidal cast iron is ductile in both tension and compression. BS EN 1563:1997) includes the properties of standard ductile cast irons.

Wrought iron

Good quality wrought iron is far more ductile than cast iron, but is a very different material to steel. Its material properties are variable, as it has a laminar microstructure that is highly dependent upon the manufacturing process used to form it. Morgan (1999) is a particularly useful source of information on wrought iron.

Microstructure

The microstructure of wrought iron, and thus its mechanical properties, are directly affected by the details of manufacture.

Wrought iron was formed by reheating cast iron to a high temperature, and stirring to remove carbon and other impurities. The cooled iron was then reworked by a combination of reheating, and squeezing, hammering and rolling. The reworking process gave a matrix of pure iron, within which the slag formed elongated stringers, aligned principally with the direction of final working. The microstructure of wrought iron depends upon the method of manufacture, and the number of times it was reworked. Repeated reworking gave finer slag stringers, which improved the strength and ductility of the wrought iron.

Material properties

The graphite fibres within wrought iron are primarily aligned in the plane of a plate, or along the axis of a rod. They therefore do not act as stress raisers under axial or in-plane tensile loading, and the material exhibits strength in tension approximating to that of the steel matrix, with considerable ductility. A typical stress-strain curve for wrought iron is shown in Figure 2.7. Wrought iron has essentially similar characteristics in tension and also compression.

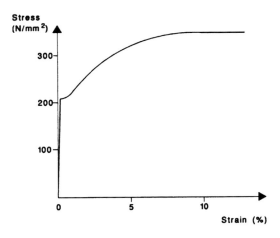

Figure 2.7   *A typical stress-strain response for wrought iron (from Bussell, 1997)*

The fatigue behaviour of wrought iron depends to some extent upon the amount of cold working carried out during the final stages of manufacture (which is difficult to assess). Cracks typically propagate from rivet holes or similar large stress concentrations.

Only very limited data is available on the fire performance of wrought iron. The available data indicates that high temperature can be beneficial to the material's strength, probably due to its high phosphorous content.

Design values for the material properties of wrought iron

No research has been carried out on wrought iron since the start of the 20th century and design remains largely based upon these tests, which are summarised by Bussell (1997). BD 21/01 (Highways Agency, 2001a) provides some guidance when assessing wrought iron structures, although it is far from comprehensive.

To some extent, wrought iron can be treated in a similar manner to steel, although it is sensible to use a higher safety factor than for modern steel. It must be recognised that wrought iron is fundamentally a different material, due to its laminar microstructure and the presence of slag inclusions.

For the vast majority of cases, wrought iron may be assumed to be a linear-elastic material. It is unlikely that wrought iron beams will have suitable geometric dimensions to allow plastic design. Where these requirements are met, the designer must be satisfied that the wrought iron is of consistently of sufficient good quality to provide the required ductility.

A wrought iron structure should always be carefully examined for laminations, inclusions and deformities. Severe corrosion, and a history of high cyclic loading, should be considered when assessing the wrought iron's remaining fatigue life.

Delamination of wrought iron

Whilst there is no prior experience with adhesive bonding to the side of a wrought iron plate, there is no theoretical reason why FRP should not be bonded to a good quality wrought iron structure, as the strength of the wrought iron perpendicular to the grain is greater than that of the adhesive.

The interlaminar strength, however, is significantly reduced by delamination, which occurs in rolled wrought iron. The iron rusts, whilst the slag does not, so that the rust tends to detach itself in flaky sheets. *It follows that adhesive bonding to a corroded wrought iron member is unlikely to be successful.* Surface delamination is easily recognised. Delamination beneath the surface can be detected using ultrasound NDT techniques.

Until the results of further research are available, strengthening wrought iron with FRP is not recommended as a general solution. Further tests must be carried out to validate its use.

2.2.4   **Carbon steel**

Carbon steel is characterised by linear-elastic behaviour up to the onset of yield, followed by considerable ductile plastic strain up to ultimate rupture, generally accompanied by strain hardening. Figure 2.8 shows a typical stress-strain response for a modern mild steel. The most suitable failure criterion for steel is the Von Mises limiting shear strain energy criterion.

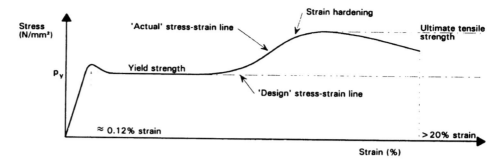

Figure 2.8   *An idealised stress-strain response for modern mild steel (from Bussell, 1997)*

Design guidance is well documented in codes such as BS 5950:2000 for building structures and BS 5400-1:1988), and BD 56/96 (Highways Agency, 1996) addresses the assessment of existing steel highway bridges.

Historic steel

Vast improvements have been made in the manufacture of steel over a number of years, and engineers should recognise that historic steel was not produced to the same standards as today. Before the production of British Standards, the shapes and sizes of steel sections were determined by each manufacturer (Bates, 1984).

It is difficult to predict the properties of historic steel unless records have been kept, and sample testing is likely to be required. In the absence of other information, BD 21/01 (Highways Agency, 2001a) recommends a characteristic strength of 230 MPa for steel produced before 1955.

Steels produced before 1922 can be of poor quality and may be laminated or include other defects. These defects should be identified during inspection of the structure, since remedial works will be required before FRP strengthening material is bonded to the surface of the metal.

2.3   IDENTIFICATION AND CHARACTERISATION OF METALLIC MATERIALS

Before designing a strengthening system, it may be necessary to carry out a series of tests to determine the properties of the materials from which the structure is made. This is particularly desirable for historic structures.

2.3.1   Date of construction

The date of construction is a good guide to the likely material of construction (see Figure 2.1). Bates (1984) gives a useful summary of the development of design methods and design codes together with the development of structural iron and steel.

From about 1880 to 1900 cast iron, wrought iron and steel were all in use. It can be difficult to distinguish between wrought iron and steel, since both were used to form riveted compound and plate girders. The only certain method to distinguish between these materials is metallographic examination of a sample sawn from the member. The sample must *not* be flame cut, as the thermal stresses induced are likely to cause member failure in brittle materials such as cast iron. Specimens must not be removed from highly stressed regions of the structure, and repair of the affected area should be considered. The orientation of the samples with respect to the member should be recorded.

2.3.2 Material properties

The properties of the metal being strengthened can often only be determined with certainty by testing samples taken from the member, noting the warnings described in the preceding section. The strength of cast iron can be approximately determined from a hardness test on the surface of the metal, as the tensile strength of cast iron is roughly correlated with its hardness (Bussell and Robinson, 1998).

There can be considerable variation in the material properties within a structure (and possibly within a member), particularly in the case of wrought iron. Several specimens (normally not fewer than 10) are required to obtain representative results. A wide spread of material properties does not necessarily indicate inaccurate test methods.

2.3.3 Structural survey

A structural survey of the members to be strengthened should be undertaken, to determine the geometry, thickness, out-of-straightness and variability of the members. The extent of corrosion and dead weight carried by member should also be assessed.

2.3.4 Connections

Particular attention should be given to connections in the structure. The strength of a connection depends upon its geometry and the ductility and strength of the components from which it is formed. The load distribution between pins, rivets or bolts depends on the accuracy of fit of the pins within the holes, and on the ductility of the pins and plates. If the pins behave in a ductile manner, a good ultimate distribution of load between bolts is developed. Potential failure modes include bolt failure in shear, bearing or flexure; and plate failure in bearing at the bolt interface, in tension or compression at the reduced section between pin holes, and in shear as a result of pin pull-out.

2.3.5 Assessing permanent stress in the structure

In most cases the structure to be strengthened will have to be assessed to determine its current carrying capacity. Detailed information on its characteristics, material properties and permanent stresses should be included in the assessment.

If the existing structure is statically determinate, the permanent stresses within it should be determined by simply assessing the dead loads acting upon it. If the structure is indeterminate, it will be necessary to assess the permanent stress in the structure before strengthening begins.

An outline strategy to assess the permanent stress within an existing structure is to:

- choose a position of the member that will provide useful information on the stresses carried by the structure, but which is not critical
- attach a strain gauge to the surface of the metallic member
- drill a hole in the structure adjacent to the strain gauge
- assess the change in stress due to the hole, and hence deduce the permanent stress in the member at the location of the hole
- repair the test area as necessary.

The change in stress can be assessed in a similar manner when samples are removed to determine material properties.

2.4   BIBLIOGRAPHY

Angus, H T (1976). *Cast iron: Physical and engineering properties*. 2nd edn, Butterworth, London

Barnfield, J R and Porter, A M (1984). "Historic buildings and fire: fire performance of cast iron structural element". *The structural engineer*, vol 62A, no 12, pp 373–380

Bates, W (1984). *Historical structural steelwork handbook*. BCSA 11/84, British Constructional Steel Association, London

Highways Agency (1996). *The assessment of steel highway bridges and structures*. BD 56/96 (DMRB vol 3, sec 4, pt 11), Stationery Office, London

Highways Agency (2001a). *The assessment of highway bridges and structures*. BD 21/01 (DMRB vol 3, sec 4, pt 3), Stationery Office, London

Bussell, M N (1997). *Appraisal of existing iron and steel structures*. P138, Steel Construction Institute, Ascot

Bussell, M N and Robinson, M J (1998). "Investigation, appraisal, and reuse, of a cast iron structural frame". *The structural engineer*, vol 76, no 3, pp 37–42

Cullimore, M S G (1967). "The fatigue strength of wrought iron after weathering in service". *The structural engineer*, vol 45, no 5, pp 193–199

Doran, D K, ed (1992). *Construction materials reference book*. Butterworth Heinemann, London

IStructE (1996). *Appraisal of existing structures*. 2nd edn, SETO, London

Morgan, J (1999). "The strength of Victorian wrought iron". *Proc Instn Civ Engrs, Structs & bldgs*, vol 134, Nov, pp 295–300

Moy, S S J (1999). "A theoretical investigation into the benefits of using carbon fibre reinforcement to increase the capacity of initially unloaded and preloaded beams and struts". Paper prepared under Link & Surface Transport Programme, Carbon fibre composites for structural upgrade and life extension – validation and design guidance. Dept of Civil & Environmental Engg, Univ Southampton

Swailes, T (1995). "19th century cast iron beams: their design, manufacture and reliability". *Proc Instn Civ Engrs, Civ engg*, vol 114, Feb, pp 25–35

Swailes, T and Marsh, J (1998). *Structural appraisal of iron-framed mills*. ICE design and practice guides, Thomas Telford, London

British Standards

BS 1452:1990. *Specification for flake graphite cast iron* [superseded]

BS 5400-1:1988. *Steel, concrete and composite bridges. General statement*

BS 5950:2000. *Structural use of steelwork in building. Code of practice for design. Rolled and welded sections*

BS EN 1561:1997. *Founding. Grey cast irons*

BS EN 1563:1997. *Founding. Spheroidal graphite cast iron*

3 Fibre-reinforced polymer (FRP) strengthening systems

An FRP strengthening system consists of two principal elements: the FRP strengthening material and the adhesive layer that bonds the FRP to the existing metallic structure. The FRP is itself a composite material composed of two constituent elements: the fibres and the binding matrix.

This chapter describes each element of an FRP strengthening system. Section 3.1 discusses the properties of fibre and matrix materials that are combined to produce an FRP composite as described in Section 3.3. The adhesive joint is described in Section 3.4. This chapter includes a discussion of the mechanical properties, thermal sensitivity, environmental degradation, creep characteristics and fatigue performance of the elements of an FRP strengthening system.

The FRP strengthening can be preformed for later bonding to a structure, or both the FRP and the adhesive joint can be formed *in situ* in a single operation. Methods of installation and manufacture are described in Section 3.2.

While this chapter provides an overview of typical FRP strengthening systems and their properties, the material supplier should always be consulted for the actual details of a specific strengthening system.

3.1 THE CONSTITUENT MATERIALS WITHIN AN FRP

When strengthening metallic structures using FRPs, it is essential to have a clear understanding of the strengthening materials being used.

The strength and stiffness of the composite are largely determined by the fibres within it. FRPs that are suitable for structural strengthening contain high-strength and high-stiffness fibres in relatively high-volume fractions. The orientation of the fibres is controlled, so that the resulting FRP is an anisotropic material. This enables high mechanical stresses to be carried safely, and allows the strengthening material to be tailored to suit the local stress patterns in the component. The in-service performance of the composite is influenced by both the fibre and the matrix material.

3.1.1 Properties of reinforcing fibres

The principal fibres used in FRP strengthening materials are based on carbon, aramid and e-glass. Indicative properties of the different fibres are given in Table 3.1.

Carbon fibres have both high strength and high stiffness. They are lightweight and have low coefficients of thermal expansion, but high electrical conductivity. Carbon fibres are available in a variety of different grades, according to the process by which they are manufactured:

- high-strength (HS)
- high-modulus (HM)
- ultra-high-modulus (UHM).

Table 3.1   *Properties of reinforcing fibres*

	Carbon fibre			Aramid fibre***	E-glass fibre
	High-strength (HS)*	High-modulus (HM)*	Ultra-high modulus (UHM)**		
Modulus of elasticity (GPa)	230–240	295–390	440–640	125–130	70–85
Strength (MPa)	4300–4900	2740–5940	2600–4020	3200–3600	2460–2580
Strain to failure (%)	1.9–2.1	0.7–1.9	0.4–0.8	2.4	3.5
Density (kg/m³)	1800	1730–1810	1910–2120	1390–1470	2600
Coefficient of thermal expansion (parallel to fibre), (10^{-6}/°C)	-0.38	-0.83	-1.1	2.1	4.9

*   Polyacryonitrile precursor (see Appendix 2)

**   Pitch precursor (see Appendix 2)

***   Aramids with the same strength but lower modulus are also available, but are not suitable for structural engineering applications.

Generally, an increase in stiffness is accompanied by a reduction in the strength and strain to failure of the fibre. UHM fibres can be very brittle and care must be exercised when handling them.

Aramid fibres are better known by the trade names Twaron and Kevlar®. Aramids have high strength and strain to failure and moderate stiffness. The longitudinal strength of the fibres is greater than that in the transverse direction, consequently aramid FRPs generally have low compressive and shear strengths.

E-Glass fibres are used for general-purpose structural applications. They are the least stiff and least strong of the fibres considered here, but are considerably cheaper than either carbon or aramid fibres.

Further details of these fibres and their manufacture can be found in Appendix 2.

To be effective, FRP strengthening material that is not prestressed must have a similar stiffness to the metallic structure being strengthened. Carbon fibres are often preferred for strengthening metallic structures since these have the highest stiffness. Aramid fibres can also be used effectively, although a greater quantity of strengthening material will be required. Unlike carbon fibres, aramid fibres are electrical insulators and may be of benefit near electrical installations (see Section 6.6.7). Glass fibre is not normally used as the longitudinal strengthening material on a member, but may be used transversely for a variety of purposes: to confine a member, to provide anchorage to the main strengthening material, to confer transverse strength to the strengthening, to act as a galvanic corrosion isolation layer between CFRP and the metal substrate (Section 6.6.7), or to act as a protective layer over CFRP strengthening material.

Fibre arrangement

The fibres are woven into yarns and fabrics, which are then incorporated in the composite material.

The fibres can be arranged to give a quasi-isotropic or an anisotropic composite material. If the fibres are arranged with equal proportions in equispaced multiple directions, for example 0°, 30°, 60°, 90°, the composite would exhibit quasi-isotropic properties. When the fibres are aligned predominately in a single direction, the composite is described as unidirectional.

The properties of composites with various fibre orientations are described in Section 3.3

3.1.2 Composite matrix material

The polymeric matrix binds and protects the fibres; it transfers force into the fibres by interfacial shear and protects the delicate fibres against aggressive environments.

The durability properties of a composite are largely dependent upon the matrix material, which determines the heat, fire and chemical resistance of the composite. An organic resin matrix is usually used in FRP composites for structural applications. The properties of a variety of resins are listed in Table 3.2.

Table 3.2 *Typical properties of resins*

	Epoxy	**Polyester**	**Phenolic**	**Polyurethane**
Modulus of elasticity (GPa)	2.6–3.8	3.1–4.6	3.0–4.0	0.5
Tensile strength (MPa)	60–85	50–75	60–80	15–25
Strain to failure (%)	1.5–8.0	1.0–2.5	1.0–1.8	10
Poisson's ratio	0.3–0.4	0.35–0.38	(not available)	0.4
Coefficient of thermal expansion (10^{-6}/°C)	30–70	30–70	80	40
Density (kg/m³)	1110–1200	1110–1250	1000–1250	1150–1200

Epoxies are generally two-part systems comprising resin and hardener components. Epoxies are available with a wide range of properties, and a variety of curing temperatures. They have high specific strengths, good dimensional stability, high temperature resistance and good resistance to solvents and alkalis, but generally have low resistance to acids. The toughness of the epoxies is superior to that of many other civil engineering polymers. They exhibit good adhesion to many substrates and low shrinkage during polymerisation. The epoxy formulation can be changed to allow the strengthening to operate at high temperatures. Epoxies are most often used as the matrix component in wet lay-up strengthening applications and as the adhesive in plate bonded strengthening.

Polyester resins are versatile, relatively low-cost (compared, for example, with epoxies), and easy to process. Polyesters are frequently used as the matrix in preformed composite components. They have moderate mechanical properties. Cure is strongly exothermic, usually accompanied by significant shrinkage and results in the release of styrene vapour. *Vinyl esters* are a subset of the polyester family, which have better thermal stability and environmental resistance, but are also characterised by significant curing shrinkage.

Phenolic resins are of particular interest for their fire-resisting properties, since they emit very little smoke and toxic fumes. They are more hygroscopic than other resins, however, and absorbed water will turn to steam and cause failure of the laminate during a fire.

Polyurethane resins are available in a wide variety of formulations. They have adequate resistance to water and high tolerance to chemicals. They are slightly weaker than epoxies and are characterised by significant curing shrinkage, creep and moisture movement.

Mechanical properties

The short-term strength and stiffness of a thermosetting polymer depend upon the degree of cross-linking. The strength characteristics depend upon the nature of the load to which the resin is subjected (for example, whether the load is a static short- or long-term loading or a dynamic or impact loading). The thermal properties of the resin also have a significant influence on the mechanical properties of the resin.

An increase in the stiffness of the resin is usually accompanied by a reduction in toughness. The toughness of a brittle, high-stiffness polymer can be increased by blending it with a tough polymer, although this reduces its adhesive strength.

Material suppliers will be able to advise on the most appropriate resin formulation for a particular application.

Thermal dependence of the mechanical properties

The mechanical properties of a polymer (and consequently of both the composite matrix and the adhesive) are temperature-dependent.

Up to a certain temperature, thermal exposure may be an advantage as it can result in a post-cure for the FRP composite and adhesive. Figure 3.1 shows the strength development for a two-part epoxy at various temperatures. As a rule of thumb, the time required for a resin or adhesive to cure halves for each 10°C rise in the temperature of the resin. For many epoxy resins, curing stops altogether below about 5°C.

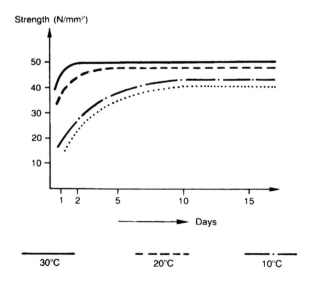

Figure 3.1 *Effect of formulation and cure temperature on the development of flexural strength of a two-part epoxy adhesive (from Mays and Hutchinson, 1992)*

Thermosetting polymers for civil engineering use are glassy in nature. As the temperature increases, two phenomena occur. First, as the polymer approaches its *glass transition temperature* (T_g), it begins to soften, as shown in Figure 3.2 for an ambient-cure epoxy. Second, at elevated temperatures, all polymers will decompose.

T_g depends upon the detailed chemical structure of the polymer. Above T_g, polymers that are not crystalline become soft elastomers or viscous liquids. If the polymer is crystalline its properties will range from a soft to a rigid solid depending upon the degree of crystallinity. Polymer composite structural units should not be exposed to temperatures above the T_g of the material. Sections 6.2.2 and 6.1.4 describe the relevance of the glass transition temperature to design, while Section 7.1.1 describes methods for determining the T_g value of a resin.

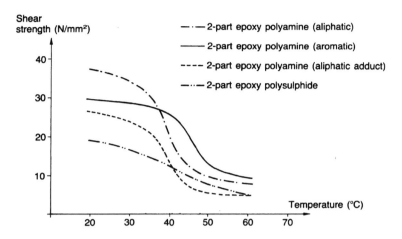

Figure 3.2   *Variation in epoxy adhesive shear strength with temperature (from Mays and Hutchinson, 1992)*

The T_g of *some* low-temperature moulded composites can be increased by further post-curing at a higher temperature, but there is a limit to the value of T_g that can be attained, irrespective of the post-cure regime. This limit varies between different resins. For example, for a typical hot-cured resin the T_g value is 5–10°C above the post-cure temperature up to a post-cure temperature of about 120°C. However, cold-cure adhesives (such as those commonly used on site in civil engineering applications) have a T_g value of around 50–55°C, which cannot be increased. Advice should be sought from the materials supplier.

As the temperature of polymers falls all polymers harden and become increasingly brittle. Their fracture toughness and critical energy release rate, which are critical properties for adhesion strength, are significantly reduced as a result. This is not usually a problem until the temperature drops to around -20°C, and then only if the polymer is exposed to this temperature for a long time. Each polymer has different low-temperature behaviour, so if an FRP strengthening scheme is expected to operate at low temperatures for long durations, the material properties must be determined from tests carried out at a representative temperature.

Figure 3.3 shows the variation in mechanical properties of an ambient cure two-part epoxy adhesive with temperature (as used to strengthen Tickford Bridge, described in Appendix 1). The results were obtained by conducting lap-shear tests (Section 7.1.1) at different temperatures. The figure shows the average shear stress in the specimen at failure, the peak shear stress (obtained by back analysis, as described in Section 5.3.1) and the mode II (shear) fracture toughness of the adhesive. The mechanical properties of the adhesive change significantly with temperature. The optimum material properties from these tests were obtained at ambient temperature (24°C). At temperatures above 40°C the properties of epoxy reduce as T_g is approached. At low temperatures (-15°C), embrittlement occurs and there is a reduction in the strength of the adhesive joint.

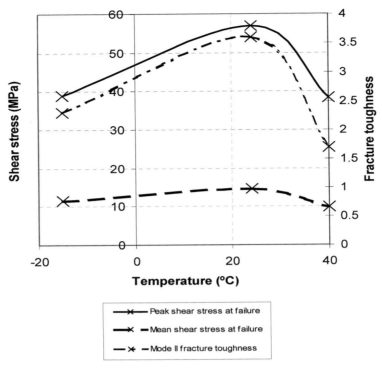

Figure 3.3   *An indication of the variation in the properties of a two-part ambient cure epoxy with temperature. (Experiments were carried out at three temperatures, with a best-fit curve indicating the approximate trend)*

Thermal properties

The *coefficient of thermal expansion* of the matrix resin (α_{matrix}) is generally considerably higher than that of conventional materials such as steel and concrete, although when fibres are introduced the net coefficient of thermal expansion is reduced (see Section 3.3.5 and Table 3.4). Differential thermal expansion must be considered in structural design (see Section 6.3.1), and is of particular importance in externally bonded strengthening of metallic structures. The coefficient of thermal expansion varies with temperature and it can be significantly different from that of the substrate.

The *thermal conductivity* of epoxies is low, so they are good heat insulators.

Creep characteristics

Polymeric matrix materials exhibit viscoelastic creep behaviour. However, FRP strengthening materials for metallic structures usually contains a high proportion of unidirectional fibres, which are not susceptible to creep. Furthermore, the permanent stress levels in externally bonded FRP strengthening are generally low enough for creep not to be significant (Section 6.1.4).

Creep is more significant in the adhesive joint, as described in Section 3.4.3

Environmental effects

The polymeric resins used in the matrix of the composite and the adhesive layer absorb water from the surrounding environment. This absorbed moisture affects their mechanical properties. The effect of an increase in water content is equivalent to a drop in the glass transition temperature (T_g) value (see Section 6.1.4). The maximum moisture absorption of a given material can be determined by the test method given in

BS EN 2378:1994. The actual long-term moisture absorption in a given environment can be predicted by Fickian moisture diffusion theory (Pierron *et al*, 2002).

Freeze-thaw cycles can result in accelerated degradation. Any trapped moisture will swell and contract, which can potentially result in delamination of the laminates, either from the steel adherend or within the FRP composite plate.

Most polymers are more chemically resistant than most metals, particularly to acids. Nevertheless, the adhesive and resin systems should be selected to suit the chemical environment. It is thus essential to establish the operating conditions under which the strengthening will function, and to seek advice from the supplier on the material's sensitivity to operating conditions.

3.2 MANUFACTURE AND INSTALLATION PROCESSES FOR DIFFERENT FORMS OF FRP

Manufacturing FRP systems can be divided into two basic approaches.

In-situ methods of manufacture, which can be:

- *manual*, including wet lay-up of laminae, with or without vacuum consolidation
- *semi-automated*, including vacuum infusion methods
- *automated*, such as filament winding.

Preformed components, which are subsequently bonded to the structure, are manufactured by processes that are typically:

- *semi-automated*, in which resin transfer methods are used to preform components
- *automated*, including pultrusion and resin transfer moulding.

The choice of strengthening material form is dictated by considerations that include:

- the extent and type of strengthening required
- the shape of the metal member
- the regularity (flatness) of the substrate
- the cross-sectional area of the reinforcement required
- the available surface area to which the reinforcement can be applied
- the interlaminar shear stresses that are expected
- site conditions
- feasibility of the installation process
- the required appearance.

3.2.1 In-situ methods of manufacture

Manual methods

Wet lay-up process

The wet lay-up process involves manually laminating the composite material using an open mould. Normally, wet lay-up techniques are used to laminate the composite material directly on to the structure. They may also be used to produce a preformed composite plate, for later bonding to the structure, but this is not generally efficient. Multiple layers of fibre mat are sequentially applied to the mould and impregnated layer by layer.

The wet lay-up process is slow and labour-intensive, since it is necessary to wait between successive layers, and there is a limit to the number of layers that can be applied in any one day. *In-situ* techniques such as wet lay-up allow FRP strengthening to conform to the profile of the substrate, and hence to be applied to curved surfaces.

Section 7.4.3 describes the installation of wet lay-up strengthening.

Pre-impregnated (prepreg) composites

Composite materials can be supplied as pre-impregnated mats, in which the fibres are encapsulated in a partially cured matrix resin (known as prepreg). The partial-cure is sufficient to give the composite integrity, but it remains pliable, allowing the composite to be applied to the structure. Prepregs are usually cured under pressure and at high temperature.

Vacuum bag consolidation

A vacuum bag technique can be used to consolidate wet lay-up, vacuum infused or prepreg composite before curing proceeds. A release film is applied over the composite, followed by a polymer membrane, which is sealed to the structure along its edges (Figure 3.4). Air is then extracted from the bag, so that a pressure of up to one atmosphere is applied to the surface of the moulding. With glass fibre, a fibre volume ratio of 55 per cent can be achieved using a vacuum bag method.

Figure 3.4   *Strengthening a curved steel beam using low-temperature moulding prepregs and vacuum bag consolidation (courtesy Taylor Woodrow, Southall, and ACG, Derbyshire, UK)*

Semi-automated methods of manufacture

Semi-automated processes are available for the application of FRPs for structural strengthening. These processes are developments of the manual methods described above. Semi-automated techniques still require manual input, but simplify the installation process and improve the quality of the finished composite.

Vacuum infusion

High-quality composites can be formed and bonded to a structure *in situ* using vacuum infusion. Various proprietary resin infusion systems exist. Dry fibre mats are stacked on the structure and overlayed by a diffusion membrane and a vacuum bag membrane. In a single operation the vacuum bag edges are sealed and a high vacuum is applied to consolidate the fibre stack (the "perform"), which draws in resin to impregnate it. The vacuum-drawn resin forms both the composite material and the adhesive bond between the FRP and the structure.

The process produces composites with a high fibre volume fraction (about 55 per cent), resulting in high strength and stiffness properties. *In-situ* vacuum infusion does have disadvantages when working overhead and it is slower and more labour-intensive than bonding preformed plates. However, it delivers better consolidation of the composite and better wet-out of the fibres than hand lamination techniques.

Low-temperature moulding prepregs

There are various formulations of low-temperature moulding composites consisting of either high-modulus or ultra-high-modulus carbon fibres pre-impregnated with an epoxy resin. These prepreg composites are cured under a combination of vacuum consolidation and elevated temperature, applied using an electric heating blanket. As for all composite strengthening systems, the degree of cure is critically dependent on the curing temperature and duration. It is therefore important to monitor the temperatures reached on the substrate side of the assembly.

The curing temperature of the matrix is typically about 65°C, giving a glass transition temperature of about 75°C. The material can be post-cured to higher temperatures.

Automated processes of manufacture

Filament winding

Filament winding is limited to the *in-situ* strengthening of axisymmetric structures such as cylindrical columns, chimneys, silos, tanks etc. Several manufacturers have developed specialised winding equipment. This comprises a carriage that travels around the circumference of the structure, laying out impregnated fibre. The impregnated fibre composite is cured at high temperature. The equipment is quite expensive and inflexible, so it is most suitable when there are many identical structures to be wrapped with FRP.

3.2.2 Preformed methods of manufacture

Manufacturing processes can be used to fabricate preformed FRP composite units, which are later bonded on to a metallic structure.

Strengthening plates and strips can be preformed by the pultrusion process, vacuum infusion or from prepregs under factory conditions. Plates can be supplied with such features as multiple fibre directions, tapered ends, a glass surface layer (to alleviate galvanic corrosion, Section 6.6.7) and peel plies (to provide a suitable surface for bonding, Section 7.3.4). The overall performance and properties of plates containing these options may vary from simple two-component composites and the actual properties need to be considered in the structural analysis.

Bonded strips or plates are normally applied in a single layer. Following preparation of the substrate surface, adhesive is spread on the plate and the substrate, and the plate pressed against the substrate to expel the air in the adhesive layer zone. The plate is held tight against the substrate during curing. This installation method has the advantage of shorter installation time, since there is less resin to be applied *in situ*. In the presence of an irregular substrate surface, a preformed plate should have sufficient stiffness to keep the fibres straight, surface variations being taken out in the adhesive thickness.

Preformed strengthening material can be prestressed prior to cure of the adhesive, allowing a transfer of permanent stress from the existing structure into the composite strengthening material. Prestress is particularly beneficial with brittle materials (such as cast iron) with low existing live load capacity (see Figure 3.5).

Semi-automated methods of manufacture

Resin transfer moulding (RTM)

In the RTM system, a fibre preform is placed on a tool or inside a mould cavity and is encapsulated inside a vacuum bag or mould cover. Thermosetting resin is injected into the mould, where it saturates the preform and fills the mould. The resin system undergoes a curing reaction to produce the finished part. The mould-filling process, which can take several minutes, is critical to obtaining a good-quality product. To prevent voids or dry spots in the finished article, the resin must fully wet-out the preform; voids cause defects that diminish the strength and quality of the cured part.

Automated methods of manufacture

Pultrusion

The pultrusion technique is a fully automated closed mould system. Continuous fibrous reinforcement rovings and mats are pulled through a reservoir of resin and a heated die. Continuous monitoring is used to ensure that the fibre placement, resin formulation, catalyst level, die temperature and pull speed are maintained at the correct levels, and to ensure that the pultruded unit has the correct appearance and properties.

Figure 3.5   *Prestressed, pultruded CFRP strengthening installed on the cast iron beams of a brick jack arch bridge (courtesy Mouchel, West Byfleet, UK)*

3.3   COMPOSITE PROPERTIES

The properties of a composite material depend upon the:

- properties of the fibre used
- properties of the matrix used
- relative proportions of the polymer and fibre (the fibre volume fraction)
- orientation of the fibres and the direction of loading
- method of manufacture.

All of these variables must be defined before material properties can be quoted.

3.3.1   Mechanical properties of a single ply composite

The *composite* properties of the fibres combined with the matrix depend, among other things, upon the level of control over the thickness of the material. In the case of pultruded plates and sheets with unidirectionally aligned fibres, the thickness is controlled relatively accurately, therefore the modulus of the composite in the fibre direction can be calculated using the *rule of mixtures*:

$$E_f = E_{fibre}V_{fibre} + E_{matrix}(1 - V_{fibre}) \qquad (3.1)$$

E_{fibre} and E_{matrix} are the elastic moduli of the fibre and matrix respectively, E_f is the elastic modulus of the composite, and V_{fibre} is the volume fraction of fibre.

In the case of reinforcement forms that are built up or impregnated *in situ*, the final thickness of the material can be rather variable, resulting in variable fibre volume fraction and composite properties. If one assumes a consistent value of volume fraction throughout the stress analysis, however, the fibre stresses and strains obtained will be correct. In this situation it can be more convenient to work in terms of the fibre volume and fibre properties and neglect the resin properties.

Composites are in general anisotropic: their mechanical properties vary according to the direction in which they are loaded. Figure 3.6 shows how the properties of a unidirectional composite vary as the direction of loading is rotated with respect to the fibre direction.

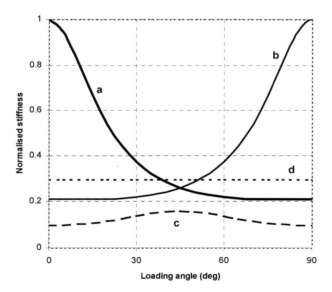

(a) unidirectional laminate, parallel to the direction of loading

(b) unidirectional laminate, normal to the direction of loading

(c) unidirectional laminate, shear stiffness

(d) stiffness of a quadriaxial laminate, with the same volume fraction.

All normalised with respect to the longitudinal stiffness (based on a glass-epoxy laminate, with fibre volume fraction of 50 per cent).

Figure 3.6   *Variation of laminate stiffness with the direction of loading*

3.3.2   Mechanical properties of a laminated composite

Layers of fibres can be stacked to produce a laminated structure, in which the fibres in different layers can be given different orientations. The fibre orientations in a composite are configured to suit the loading envelope that the material is required to carry. It is unusual to use a purely unidirectional composite, since transverse fibres are required to give the composite sufficient transverse and shear strength to cater for secondary stresses and prevent cracking. Nevertheless, pultruded plates and strips for strengthening applications generally have a high proportion of unidirectional fibre, since applied stresses are prevalently longitudinal. The properties of a composite built up from a stack of fibre mats orientated in different directions can be predicted by laminate analysis theory (see Jones, 1998).

A quadriaxial fibre layup having equal proportions of fibres in the 0°, 90°, +45°, -45° directions produces a composite with quasi-isotropic properties. Figure 3.6 compares the stiffness of a quadriaxial composite (line d) with a unidirectional composite, having the same volume of fibres. The higher stiffness of the unidirectional fibre is apparent, but the orthotropic nature of a quadriaxial composite may be required where the applied stresses are highly multidirectional. This could be the case if the laminate needs to incorporate a hole, which would create a multidirectional stress field in its vicinity.

Once the composite has been manufactured, it is advisable to conduct tests on the material to determine its actual properties. For preformed strengthening systems, the manufacturer usually undertakes the tests.

3.3.3 Failure criteria

Composite materials reinforced with long fibres generally have an almost linear stress-strain curve up to failure. Several failure criteria are available to describe ultimate failure of the composite. The maximum strain and Tsai-Hill criteria are most commonly applied to composite strengthening material (Jones, 1998).

Maximum strain failure criterion

The maximum strain criterion is most suitable for *monoaxial* stress states, although it is also used for multiaxial stress states. The limiting strain ($\overline{\varepsilon}_{xf}, \overline{\varepsilon}_{yf}$ or $\overline{\gamma}_{xyf}$) depends upon the direction in which the load is applied (with respect to the fibre arrangement within the composite).

$$\varepsilon_{xf} < \overline{\varepsilon}_{xf}, \quad \varepsilon_{yf} < \overline{\varepsilon}_{yf}, \quad \gamma_{xyf} < \overline{\gamma}_{xyf} \tag{3.2}$$

The above requirements apply simultaneously.

Tsai-Hill failure criterion

The Tsai-Hill criterion is suitable for *multiaxial in-plane* loading when the applied stresses σ_{xf}, σ_{yf} are both tensile or both compressive.

$$\left(\frac{\sigma_{xf}}{\overline{\sigma}_{xf}}\right)^2 - \left(\frac{\sigma_{xf}\sigma_{yf}}{\overline{\sigma}_{xf}^2}\right) + \left(\frac{\sigma_{yf}}{\overline{\sigma}_{yf}}\right)^2 + \left(\frac{\tau_{xyf}}{\overline{\tau}_{xyf}}\right)^2 \leq 1 \tag{3.3}$$

where ($\sigma_{xf}, \sigma_{yf}, \tau_{xyf}$) are the stresses in the composite, and ($\overline{\sigma}_{xf}, \overline{\sigma}_{yf}, \overline{\tau}_{xyf}$) are the corresponding strength parameters. The *x* axis corresponds to the principal fibre direction.

3.3.4 Influence of the method of installation on the strengthening properties

The manufacturing process and method of installation affect the final properties of the strengthening material, due to the different levels of variability and defects that they can achieve. The manufacturing process affects the:

- thickness and volume fraction of the laminate
- orientation of the fibres
- straightness of the fibres
- temperature and duration of the cure cycle
- incidence of fibre breakages, voids, fibre disbands, unplanned fibre folds and other defects
- the glass transition temperature.

Manual fabrication methods, such as wet lay-up, tend to give lower values of strength and stiffness compared with semi-automated processes (such as pultrusion), due to the higher curing temperature, degree of compaction, wet-out, and dimensional accuracy of the latter techniques. Table 3.3 indicates the effect of manufacturing process upon the properties of the composite.

Table 3.3 *Effect of manufacturing process on typical UHM carbon laminate (from Moy, 2001b)*

Process	Fibre volume fraction	Void content	Stiffness parallel to the principal fibre direction (GPa)
Hand lay-up	0.4	Up to 5%	230
Vacuum infusion	0.54	< 1%	310
Prepreg	0.60	1–2%	360

In general, the higher the temperature and the longer the duration of curing, the better the strength, toughness, and glass transition temperature of the resin. The void content depends upon the resin viscosity and the method of consolidation used. For *in-situ* techniques, such as vacuum infusion and wet lay-up, if the substrate surface is irregular the fibres will be forced to follow the surface. They will be weakened as a result, and there will be a tendency for the composite to debond when the fibres are loaded because of the tendency of the fibres to straighten under load.

3.3.5 Coefficient of thermal expansion

The coefficient of thermal expansion of an FRP composite depends both upon the materials used and upon the lay-up of the composite. Values of the coefficient of thermal expansion for specific composite products should be provided by the manufacturer (see Appendix 3 for example).

Table 3.4 gives values of the coefficient of thermal expansion for typical composite arrangements and contrasts these with the coefficient of thermal expansion of the metallic substrate. Note in particular that carbon fibres *contract* when heated, giving a negative value of the coefficient of thermal expansion (although the resin reduces this effect). There can thus be a large difference between the coefficient of thermal expansion of the metal and that of the strengthening material.

Table 3.4   *Typical coefficients of thermal expansion (FRP values from Moy, 2001b)*

Material	Arrangement	Coefficient of thermal expansion ($\times 10^{-6}\mathrm{K}^{-1}$)	
		0° axis	90° axis
Glass FRP	Unidirectional (0°)	8.4	20.91
	Quadraxial (0°, + 45°, - 45°, 90°)	11.0	11.0
HS carbon FRP	Unidirectional (0°)	0.52	25.5
	Quadraxial (0°, + 45°, - 45°, 90°)	9.76	9.76
UHM carbon FRP	Unidirectional (0°)	-0.07	25.68
	Quadraxial (0°, + 45°, - 45°, 90°)	9.6	9.6
Resin (typical value)		60	
Mild steel		12	
Cast iron		11	

Where the coefficient of thermal expansion of a composite is not supplied by the manufacturer, it can be estimated from the properties of the fibres (α_{fibre}, E_{fibre}, ν_{fibre}, V_{fibre}) and matrix (α_{matrix}, E_{matrix}, ν_{matrix}), where α is the coefficient of thermal expansion, E the modulus of elasticity, ν the Poisson's ratio, V the volume fraction using:

$$\alpha_{\text{longitudinal}} = \frac{\alpha_{\text{matrix}}\left(1 - V_{fibre}\right)E_{\text{matrix}} + \alpha_{\text{fibre}}V_{\text{fibre}}\,E_{\text{fibre}}}{\left(1 - V_{fibre}\right)E_{\text{matrix}} + V_{\text{fibre}}\,E_{\text{fibre}}} \tag{3.4}$$

$$\begin{aligned}\alpha_{\text{transverse}} = &\left(1 + \nu_{\text{matrix}}\right)\alpha_{\text{matrix}}\left(1 - V_{fibre}\right) + \left(1 + \nu_{\text{fibre}}\right)\alpha_{\text{fibre}}V_{\text{fibre}} \\ &- \alpha_{\text{longitudinal}}\left(\nu_{\text{fibre}}V_{\text{fibre}} + \nu_{\text{matrix}}\left(1 - V_{fibre}\right)\right)\end{aligned} \tag{3.5}$$

The equation for the transverse coefficient of thermal expansion is only a first approximation and tests are recommended to determine this value if it is critical to the design (Section 7.1.1).

3.3.6   Creep and relaxation

The composite creep behaviour of a fibre-reinforced polymer under permanent load results from the progressive transfer of stress from the matrix to the fibres due to creep of the resin. Since the stiffness of a composite in the principal fibre direction is governed by the fibres, a composite with a high volume fraction fibres will have relatively low creep (see Section 3.1.2).

3.3.7   Durability

Durability of a material or structure is generally defined as the length of time that it performs satisfactorily in terms of cracking, oxidation, chemical degradation, delamination and wear under specified loading and environmental conditions, and given appropriate maintenance.

Composites in the civil construction and shipbuilding fields have a track record of about 30 years. They have been found to be generally durable when correctly designed, made and maintained. Durability has generally been assessed on the basis of accelerated ageing tests, since mathematical durability models have not been reliable (Griffith, 1978).

3.3.8 Fatigue resistance

The fatigue behaviour of fibrous composite materials is more complex than that of metals, due to the multiplicity of damage modes that can be grown by fatigue cycles and their possible interaction. However, the magnitude of the peak stress in a load cycle is usually a small proportion of the corresponding ultimate stress in practical applications, hence the in-plane fatigue endurance of composites is generally good. The most fatigue-sensitive modes of failure are matrix-dominated modes, such as debonding of the adhesive layer. Adequate factors of safety for these modes, such as those given in Section 6.1.4, will normally ensure sufficient fatigue endurance of the adhesive.

The potential damage modes include:

- fibre fracture
- matrix cracking
- fibre-matrix interface failure
- delamination.

Appendix 4 describes fatigue in composites in more detail.

3.3.9 Fire performance

Contrary to common perceptions, the correct choice of fibre and matrix materials results in a composite that is good at resisting fire. Composites are used in ships and offshore platforms, which are subject to stringent fire regulations (Gibson, 1998). Greater thicknesses of composite are used in these applications to meet specific fire performance requirements. The fire performance of composites has been demonstrated by considerable research for the offshore oil and gas production industry.

Composites have considerably lower thermal conductivities than metals. They absorb heat energy by ablation processes and the surface combustion zone acts as protective layer, delaying the spread of heat through the material. They contain a high proportion of fibres and fillers that are incombustible.

The most important composite fire performance issues are:

- spread of flame
- fire endurance (ability to sustain load during a fire)
- the production of smoke, which reduces visibility and can be toxic
- release of heat
- protection of inner face from excessive temperature rise.

Two limit temperatures define the response of a polymer with increasing temperatures:

- the glass transition temperature, T_g (Section 3.1.2)
- the temperature of chemical degradation (which may be 300–400°C, but depends upon the polymer).

If a composite material is heated to a temperature above T_g, the modulus and strength of the composite are reduced (Section 3.1.2). Provided the temperature of chemical degradation is not exceeded, the loss of modulus is reversible. Above the chemical degradation temperature, thermal damage results in an irreversible loss of the material's load-bearing characteristics.

There are some polymers that do not burn easily. For example, composites having a phenolic resin matrix do not combust easily and have virtually no smoke emission. The fire performance of resins can be greatly improved by the incorporation of fire-retardant fillers. However, the improved fire performance of composites with a phenolic matrix and of those containing fire-retardant fillers may come at the expense of worse mechanical or durability characteristics.

3.4   STRUCTURAL ADHESIVE JOINTS

Where the composite strengthening material is formed *in situ*, the matrix resin also bonds the strengthening component to the existing member.

Preformed composite components, however, must be applied to the existing structure using a separate structural adhesive. The adhesive is usually applied to the composite strengthening and the existing structure as a paste. Film adhesives are widely used in the aerospace sector to apply composite patches to aluminium (Baker, 1996), but cannot fill any imperfections in the substrate. More rigorous surface preparation will be required if film adhesives are to be used for civil engineering applications.

Possible resin types for use in the composite matrix are described in Section 3.1.2, and the same selection of adhesive types can be used to bond preformed strengthening components to a metallic member. Structural adhesives will generally have a lower cure temperature, as they must be applied under site conditions, and therefore a lower T_g than the matrix resin. Epoxy resins are most often used as adhesives, but material suppliers will advise on the most appropriate adhesive for a particular application.

Several different types of epoxy bonding adhesives are available for structural strengthening, and a summary of those available in the UK are listed in Appendix 3. Single component epoxies have ample pot life, but require raised temperatures to cure fully (Section 7.5). Two component epoxy adhesives will cure at ambient temperature after 16–24 hours. Since the pot life of the adhesive is limited to 20–45 minutes, application to large areas must be carefully planned and implemented.

The properties of structural adhesives were discussed in Section 3.1.2. The behaviour of an adhesive joint depends not only upon the properties of the adhesive, but also upon the interfaces between the adhesive and the two adherends (the composite and the metal). The behaviour of a joint is greatly influenced by the presence of bond defects at the interfaces, such as voids in the adhesive or contamination of the interfacial surfaces.

3.4.1   Durability of an adhesive joint

The most common and most important factor influencing the long-term behaviour of unprotected adhesive joints to metal substrates is the presence of high humidity or the ingress of moisture. The moisture may enter and affect adhesively bonded joints by:

- diffusion through the adhesive
- transport along the interface
- capillary action through cracks and crazes in the adhesive
- passing through defects of holes in the substrate.

Poor preparation of the metallic substrate or the CFRP leaves the adhesive joint vulnerable to environmental degradation, which may lead to failure close to the interface between the adhesive and the metal adherend.

The presence of water within the adhesive and at the adhesive-substrate interface may weaken the joint by:

- altering the adhesive properties in a reversible manner, such as by plasticisation
- altering the adhesive properties in an irreversible manner by causing it to hydrolyse, crack or craze
- attacking the interface, either by displacing the adhesive or by corrosion of the metal substrate.

Some adhesives are more susceptible to moisture ingress than others, and different types of epoxies have differing resistances. An adhesive with a high degree of resistance to moisture absorption must be selected. The adhesive joint must also resist chloride attack, and degradation due to freeze-thaw cycles (Section 3.1.2). The materials supplier will advise on the correct choice of adhesive.

3.4.2 Fatigue resistance of an adhesive joint

The adhesive joint is more susceptible to fatigue failure than the FRP strengthening material. An adhesive joint can have superior fatigue characteristics to an equivalent riveted joint, however.

The performance of a joint subjected to fatigue is dependent on its geometric configuration. Regions containing stress concentrations (Section 5.3) are most susceptible to fatigue damage.

There have been a few fatigue tests on FRP-strengthened structures. Small-scale fatigue tests (Mertz *et al*, 2001 and Miller *et al*, 2001) using different adhesives to apply double reinforcement to a steel substrate showed no sign of debonding. These were followed by fatigue tests on full-size girders (with a high degree of initial corrosion), which again showed no sign of fatigue damage or loss in stiffness after 10 million cycles of cyclic loading (taken as representative of in-service conditions). On the basis of these tests, Miller *et al* (2001) predicted that the fatigue life of the FRP strengthening would exceed that of the existing highway bridge to which it had been applied.

3.4.3 Creep characteristics

Polymeric adhesives exhibit viscoelastic behaviour. The operating temperatures of many of these materials coincide with, or are close to, their viscoelastic phase, so creep can be significant where the adhesive is subjected to permanent loading. Under sustained loading, continued creep will eventually lead to creep rupture, which places a limit on the life of the adhesive.

In the majority of FRP strengthening schemes for metallic structures, the *permanent* load carried by the strengthening is low. If load-relief jacking (Section 4.2.2) is used to transfer dead loads into the FRP strengthening, the permanent load carried across the adhesive joint can be significant. Similarly, permanent adhesive stress might be present if the FRP is prestressed (Section 4.2.1).

The magnitude of creep deformation depends on the load history, as well as on the temperature and moisture environment. To minimise creep, it is important to ensure that the service temperature does not approach the glass transition temperature of the polymer (Section 6.2.3).

The empirical Findley Power Law has been adopted and recommended by the ASCE *Structural plastics design manual* (ASCE, 1984) for the long-term analysis and design of fibre-reinforced polymer sections, and has been used extensively to model the creep behaviour of fibre-reinforced polymers, with good agreement with experimental results (Bank and Mosallam, 1992). Findley's Law is given in the design chapter (Section 6.1.4).

3.4.4 Fire performance of adhesives

It is often the adhesive that governs the fire performance of a composite strengthening scheme. The structurally effective adhesives that produce the fewest toxins upon decomposition are acrylics and some epoxies. Toughened acrylic adhesives have been shown to perform well for steel plate bonding, where the adhesive must not soften excessively under a hot surface and can meet the requirements of BS 476-7:1997.

Two-part cold-cured epoxies are generally likely to offer the best resistance to fire, since they have the following characteristics:

- stability at relatively high temperatures
- self-extinguishing after removal of the source of ignition
- absence of potential cyanide-based by-products (such as those produced during decomposition of the urethane component of polyurethane adhesives)
- do not contain significant quantities of chlorine or sulphur.

However, these should be used with caution as epoxy resins are commonly cured using nitrogen-containing hardeners. The material suppliers should be consulted as to whether harmful by-products are produced when the adhesive burns.

3.5 BIBLIOGRAPHY

ASCE (1984). *Structural plastics design manual*. ASCE Manuals & Reports on Engineering Practice No 63, American Society of Civil Engineers, Reston, VA

Bank, L C and Mosallam, A S (1992). "Creep and fatigue of a full-size fibre-reinforced plastic pultruded frame". *Comp engg*, vol 2, no 3, pp 213–227

Bowditch, M R (1996). "The durability of adhesive joints in the presence of water". *Int j adhesion and adhesives*, vol 16, no 2, pp 73–79

Brinson, H P, Morris, D H and Yeow, Y T (1978). "A new experiment method for accelerating characterisation of composite materials". *Proc 6th int conf experimental stress analysis, 18–22 Sep, Munich*, p 395

Concrete Society (2000). *Design guidance for strengthening concrete structures using fibre composite materials*. Technical Report 55, Concrete Society, Crowthorne

Dao, M and Asaro, R J (1999). "A study on failure prediction and design criteria for fiber composites under fire degradation". *Composites, Part A*, vol 30, pp 123–131

Davies, J M, Dewhurst, D, McNicholas, J B and Wang, H-B (1995). "Fire resistance by test and calculation". In: *Proc int conf fire safety by design, Jul, Univ of Sunderland*

Demers, C (1998). "Tension-tension axial fatigue of E-glass fiber-reinforced polymeric composites. Fatigue life diagram". *Const and bldg materials*, vol 12, no 1, pp 303–310

Findley, W N (1971). "Combined stress creep of non-linear viscoelastic material". In: A L Smith and A M Nicolson (eds), *Advances in creep design*. Applied Science Publications, London

Gibson, A G, Wu, Y S, Chandler, H W, Wilcox, J A D and Bettess, P (1995). "A model for the thermal performance of thick composite laminates in hydrocarbon fires, composite materials in the petroleum industry". *Revue de l'Institute Francais du Petrole* (special issue), vol 50, no 1, pp 60–74

Griffith, W I, Morris, D H and Brinson, H F (1978). *The accelerated characterisation of composite materials*. VPI engineering series VPI-E-78-3. Virginia Polytechnic Institute College of Engineering, Blacksburg, VA

Hahn, H (1976). "Fatigue behaviour of composite laminates". *J composite materials*, vol 10, pp 156–180 and pp 266–278

Henderson, J B and Wiecek, T E (1987). "A mathematical model to predict the thermal response of decomposing expanding polymer composites". *J composite materials*, vol 21, pp 373–393

Hertzberg, R W (1996). *Deformation and fracture mechanics of engineering materials*. 4th edn, John Wiley & Sons, pp 201–375 and pp 521–589

Hollaway, L C and Head, P R (2001). *Advanced polymer composites and polymers in the civil infrastructure*. Elsevier Science

Hull, D and Clyne, T W (1996). *An introduction to composite materials*. 2nd edn, Cambridge University Press

Hutchinson, A R (1997). *Joining of fibre-reinforced polymer composite materials*. Project Report 46, CIRIA, London

IStructE (1999). *A guide to the structural use of adhesives*. SETO, London

Jones, R M (1998). *Mechanics of composite materials*. 2nd edn, Taylor & Francis

Kinloch, A J (1983). *Durability of structural adhesives*. Elsevier Applied Science Pub, Amsterdam, ch 1

Lee, S M (1990). *International encyclopaedia of composites*. John Wiley & Sons, vol 2, pp 107–111

Mandell, J. (1982). "Fatigue behaviour in fibre resin composites". In: G Pritchard (ed), *Developments in reinforced plastics*. Kluwer, Dordrecht, vol 2, pp 67–107

Mays, G C and Hutchinson, A R (1992). *Adhesives in civil engineering*. Cambridge University Press

Mertz, D R, Gillespie, J W, Chajes, M J and Sabol S A, (2001). *The rehabilitation of steel bridge girders using advanced composite materials*. IDEA program final report, contract NCHRP-98-ID051, Transportation Research Board, National Research Council

Miller, T C (2000). "The rehabilitation of steel bridge girders using advanced composite materials". Master's thesis, University of Delaware, Newark

Miller, T C, Chajes, M J, Mertz, D R and Hastings, J N (2001). "Strengthening of a steel bridge girder using CFRP plates". In: *Proc New York City bridge conf, 29–30 Oct, New York*

Moy, S S J, ed (2001b). *FRP composites – life extension and strengthening of metallic structures*. ICE design and practice guide, Thomas Telford, London

Pierron, F, Poitette, Y and Vautrin, A (2002). "A novel procedure for identification of 3D moisture diffusion parameters on thick composites, theory, validation and experimental results". *J composite materials*, vol 36, no 19, pp 2219–2244

Reifsnider, K L (1982). "Analysis of fatigue damage in composite laminates". *Int j fatigue*, vol 2, no 1, pp 3–11

Rotem, A (1993). "Load frequency effect on fatigue strength isotropic laminates". *Composites science and technology*, vol 46, no 2, pp 129–139

Tuttle, M E, Mescher, A M and Potocki, M L (1997). "Mechanics of polymeric composites exposed to a constant heat flux". In: *Proc ASME intl mech engg conf: Composites and functionally graded materials*. MD-Vol 80, ASME, Dallas

British Standards

BS 476-7:1997. *Fire tests on building materials and structures. Method of test to determine the classification of the surface spread of flame of products*

BS EN 2378:1994. *Fibre reinforced plastics. Determination of water absorption by immersion*

4   Conceptual design

> Section 4.1 describes appropriate uses of FRP materials to increase the capacity of a structural member, increase its stiffness, or to extend its fatigue life.
>
> Of particular importance for cast iron structures is the permanent stress in the structure, and Section 4.2 describes methods for transferring this permanent stress into the strengthening material by using prestressed FRP, or load-relief jacking. The selection of a strengthening scheme is discussed in Section 4.3. Section 4.4 describes conceptual design issues for cast iron, wrought iron and steel structures, in a variety of structural forms.

4.1   FORMS OF EXTERNALLY BONDED FRP STRENGTHENING SYSTEMS FOR METALLIC STRUCTURES

4.1.1   FRP strengthening schemes for different structural requirements

A metallic structure may require strengthening due to one (or more) of the following structural deficiencies, illustrated in Figure 4.1:

(a) capacity in axial or flexural tension

(b) capacity in flexural shear

(c) compressive bearing capacity

(d) insufficient stiffness, including
     i.   excessive deflection
     ii.   inadequate buckling capacity
     iii.   excessive dynamic response

(e) a steel structure may be approaching the end of its fatigue life

(f) connection capacity.

Strengthening for axial or flexural tension

FRP strengthening material is most effective when used to strengthen a member in axial or flexural tension (Figure 4.1a). The strengthening element is bonded to the tensile face of the member, with its ends in regions of low stress, so that anchorage of the ends of the strengthening element is not critical.

Possible failure modes of a strengthened section include tensile rupture of the substrate, tensile rupture of the FRP, delamination of the FRP from the substrate, and long-term fatigue or creep rupture failure.

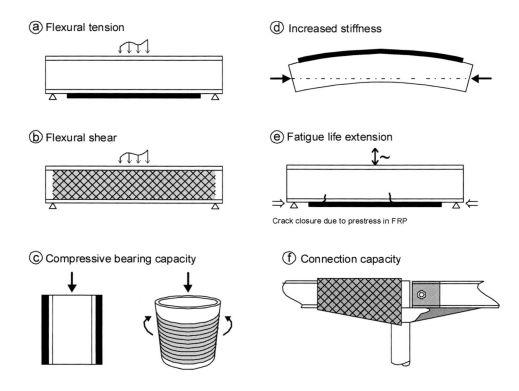

Figure 4.1 *Types of strengthening*

Strengthening for flexural shear

Externally bonded FRP strengthening is less effective for strengthening a metallic beam in flexure-shear than in flexural tension. Quasi-isotropic or a ± 45° fibre arrangement is required (Figure 4.1b).

Anchorage at the edge of the FRP strengthening component is a more critical issue, since the edges of the strengthening plate are close to the zones of maximum stress. For a thin-walled flanged section, the shear stress remains close to its maximum value right up to the junction of the web with the flange. For shear stress to be transferred into the strengthening material, the FRP must be secured with auxiliary mechanical anchorages, such as bolted anchor strips.

Strengthening for flexural shear is not underpinned by as much research as strengthening for axial or flexural tension. A test programme to back up any such scheme is recommended.

Strengthening for compressive bearing capacity

FRP strengthening material can be bonded to a compression element of a member to increase its compressive capacity, in much the same way that it is used to increase a member's tensile capacity. FRP is normally much weaker in compression than tension, however, and failure of the FRP in compression is likely to be by microbuckling and delamination.

With a concrete columns, fibre is applied in the circumferential direction to increase its compressive capacity. Similarly, a brittle compressive member, such as a circular cast-iron column, might be strengthened by circumferential fibres. These fibres confine the cast iron and prevent the localised cracking that would ensue upon failure of the member. As no research has been carried out on strengthening metallic columns with circumferential FRP, no guidance is given in this report.

Increasing the stiffness of a member

Increasing the buckling capacity, bearing strength, and stiffness of a structure of a metallic member generally requires a significant increase in the flexural stiffness of the plate or section. Bonded FRP strengthening material can be used to provide this increase in flexural stiffness (Figure 4.1d). It is particularly effective for very thin-walled structures such as profiled steel sheeting and cold-formed sections. The stiffness of thicker plates and members can also be increased effectively using FRP.

It is often preferable to bond the FRP strengthening material over the whole member, especially where a visually unobtrusive, non-invasive repair is required. Greater structural efficiency can be achieved by applying FRP material to form stiffeners or bracing elements. The local buckling resistance of the FRP stiffeners should be checked.

Extending the fatigue life of a structure

Externally bonded FRP can be used to reduce the live load stress within a metallic member and so increase its fatigue life. The fatigue life of a metallic member can be further increased by prestressing the FRP material (Section 4.2), as this places the substrate in compression and partially closes any fatigue cracks (Figure 4.1e) (Bassetti, 2001; Bassetti *et al*, 2000a, 2000b). This technique is most relevant for steel structures. A cracked section should not generally be allowed in cast iron, as discussed in Section 5.2.5.

Strengthening connections

The connections are often the weakest part in a structure and the most difficult details to strengthen, due to greater geometric irregularity and complex stress distributions (see Section 4.4).

4.1.2   **FRP strengthening for different types of member**

Beams

Beams commonly require strengthening for flexural capacity, shear capacity or bearing capacity. The flexural capacity of a material that is weaker in tension than in compression is improved by applying FRP plates to the tension face of the beam. The FRP should be extended to near a point of zero moment, to minimise the stresses in the adhesive joint. If the reinforcement takes the form of multiple layers, these should be curtailed in accordance with the stress profile.

Struts

The loading on a compression strut will normally consist of an axial compression combined with a bending moment. The critical failure mode can be yielding, buckling or cracking. If the critical failure mode is yield, then plate reinforcement can be added to control the compressive or tensile stresses, although, as already stated, FRP is not efficient in compression. If buckling is the critical failure mode, then it may be more efficient to bond stiffeners than flat plates. If cracking of the substrate governs, FRP material can be bonded to the member to carry a proportion of the applied load. However, the FRP is unlikely to be as efficient as with a ductile substrate, as its full load capacity cannot be developed.

Arches

Arches are curved compression members that are also required to carry significant bending moments, depending on the material used in their construction. In metallic arches, the behaviour of longitudinal strengthening plates is similar in all respects to that of a beam or a beam-strut, except for one additional, and potentially critical, aspect. When the strengthening plates are stressed longitudinally, through thickness peel stresses are induced as a result of the curvature of the plates.

Ties

Strengthening ties is generally relatively straightforward, except at the member ends where the load has to be transferred from the reinforcing plate into the element loading the tie.

4.2 TRANSFER OF PERMANENT STRESS INTO THE STRENGTHENING

When a strengthening element is installed the structure will be carrying permanent loads, such as dead-weight. If the strengthening material is simply bonded to the structure, none of these permanent loads will be transferred into the strengthening material (as shown in Figure 4.2a). The strengthening will only carry live loads applied to the structure after it has been strengthened.

This is not usually of concern with ductile wrought iron and steel structures, in which load is redistributed to the FRP as the metal yields. Cast iron, however, is brittle, and may have low residual strain capacity, upon which the effectiveness of the strengthening material critically depends.

Two methods can be used to transfer a portion of the permanent stress from the metallic structure into the FRP:

(a) prestressing the strengthening component

(b) load-relief jacking.

Both prestressed FRP and load-relief jacking are more complex to install than a simple unstressed strengthening scheme. Moreover, they result in higher permanent stresses in the adhesive joint, which are potentially susceptible to creep (see Section 6.1.4). Compared with the large quantities of unstressed UHM CFRP otherwise required, however, prestressed FRP and load-relief jacking may be economically favourable. The installation requirements are discussed in Section 7.6.

4.2.1 Prestressed FRP strengthening

A prestressed FRP plate (Figure 4.2b) can be used to carry a proportion of the permanent stress. As prestressing results in high adhesive stresses at the ends of the bonded plate, it is usually necessary to provide a mechanical permanent anchorage (Section 6.6.2) at both ends of the strengthening to transfer the prestress into the existing member.

The installation costs of prestressed FRP are greater than those for a simple, unstressed strengthening scheme. There is a greater risk of damaging the FRP during installation, due to accidental overload, stress concentrations in the tab zone and failure to maintain the correct alignment of the end tabs during stressing.

(a) Unstressed strengthening

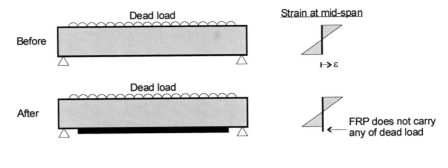

(b) Prestressed strengthening

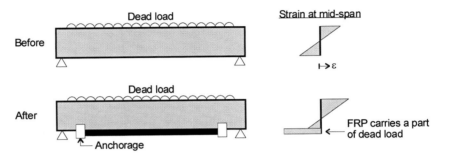

(c) Load-relief jacking

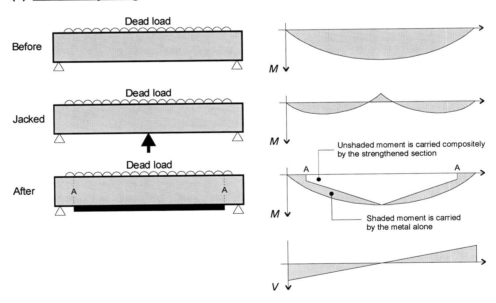

Figure 4.2 *Transfer of permanent load into the strengthening material*

4.2.2   Load-relief jacking

An alternative to prestressing is load-relief jacking. The structure is first jacked to relieve it of part of its dead load, and the strengthening material is bonded to the structure in this state. When the jacks are removed, the strengthening component becomes stressed and carries a proportion of the permanent load. The amount of permanent load transferred into the strengthening component depends upon the stiffness of the strengthening relative to that of the unstrengthened structure. It can be varied to some extent by choosing the positions at which the jacks are applied and the magnitude of the jacking loads.

As shown in Figure 4.2c, the contribution of the strengthening component to carrying the permanent load is greatest at mid-span and reduces to zero near the ends of the span. This usually means that expensive anchorage details are not required with load-relief jacking. However, the adhesive joint carries shear stresses due to the permanent dead loads (see Figure 4.2c), and these shear stresses exist along the full length of the FRP. These long-term, permanent stresses should be taken into account in the design.

Section 7.6.2 describes installation issues relating to load-relief jacking.

When strengthening a cast-iron member care should be taken not to exceed the tensile strength of the cast-iron in the temporary jacking stage. This may limit the amount of preload that can be applied.

4.3   THE SELECTION OF A STRENGTHENING SOLUTION

The type and form of strengthening material, the layout of the strengthening works, and the installation method can all be selected for a particular strengthening scheme. This means that there are various strengthening options available for metallic structures. The suitability of an externally bonded FRP strengthening solution must be assessed against conventional strengthening solutions, and strategies that allow continued use of the structure without strengthening.

Section 9.1 gives a summary of the alternatives to FRP strengthening.

4.3.1   Selection criteria

The installed cost of a strengthening scheme is usually dominated by the cost of installation and disruption. FRP composites can be installed rapidly and with little disruption to the structure's users, so the economic case for selecting FRP composites over other strengthening methods is usually self-evident (Section 9.1). The principal selection criteria reduce to questions of technical performance and the ability to meet the design requirements. These include:

- how may the requisite strength and stiffness improvements be most effectively achieved?
- how may the required quality and reliability be most effectively achieved?
- is the long-term performance of the strengthened structure likely to be satisfactory?

Factors to be considered when selecting a strengthening solution include:

- *characteristics of structure to be strengthened:* type of material, existing permanent stress level and direction, existing load capacity, required increase in load capacity, shape, surface condition, state of corrosion (Chapter 2)
- *material costs*
- *installation costs* (Chapter 7): strengthening with restricted access while maintaining service is highly labour-intensive
- *health and safety implications* (Section 7.2)
- *direct disruption costs:* loss of revenue while the facility is out of service (for example, transportation networks, factories, businesses, process plants)
- *indirect disruption costs:* cost of delay while the facility is out of service
- *access to the structure:* FRP materials can be carried through small spaces and can be applied behind existing services
- *cost/difficulty of maintenance* (Chapter 8)
- *targeting the strengthening at the weakness:* when a weak zone is difficult to access it may be feasible to strengthen an alternative location in order to relieve stress in the weak area (for statically indeterminate structures only)
- *geometric constraints*: such as headroom requirements beneath the member to be strengthened
- *minimal intervention:* this is particularly important for structures of historical significance, where the impact of the strengthening scheme must be minimal
- *environmental:* including the eventual method of decommissioning (Section 9.4).

4.3.2 Maximum amounts of FRP strengthening

There is no limit in principle to the quantity of FRP strengthening material that can be applied to a metallic structure. The quantity of FRP is limited only by the particular circumstances, for example, by the strength of the adhesive joint or headroom requirements beneath a bridge. However, a large amount of FRP strengthening material has cost implications, and there will be a point at which alternative strengthening solutions or replacement may become more attractive options than strengthening.

The ability of a structure to carry load in the event of significant damage to the FRP should be considered. The consequences of damage to the FRP should be subject to a risk assessment, and this may limit the amount of strengthening that can be applied.

Where there is marginal difference in benefit between an FRP strengthening solution and an alternative option, a more sophisticated analysis (such as finite element modelling) should be undertaken to determine a less conservative strengthening demand. This may, for example, justify the use of FRP strengthening materials, rather than demolishing and rebuilding.

4.4 CONCEPTUAL DESIGN FOR DIFFERENT SUBSTRATE MATERIALS

4.4.1 Cast iron structures

Cast iron sections are generally massive compared with equivalent steel structures. Thus, the addition of a thin layer of FRP may make little difference to the elastic stiffness of the section. Moreover, due to the relatively low cracking strain of the cast iron, the FRP needs to have as high a modulus as possible to be effective in strengthening the section. Consequently, UHM CFRP is frequently found to be the most economic form of passive strengthening for cast iron structures.

The behaviour of a brittle material strengthened with externally bonded FRP is critically dependent on the permanent stress in the metal at the time the FRP is applied, and it is important to take this (and any applied prestress in the composite and substrate) into account in the analysis. Where tests are carried out to characterise the metallic material and adhesive interface, it is equally important to preload the metallic member being tested to the same level of stress as expected in the field before the FRP strengthening material is applied.

Cast iron arches

Most elements of cast iron arch bridges are curved in one plane. These range from the primary arch ribs to spandrel rings. Many cast iron arch bridges incorporate multiple arch frames and transverse trusses to distribute live loads between the arches. Although the arches were designed to carry predominantly axial compression forces, in reality they need to be able to withstand significant bending moments due to statically indeterminate frame action, foundation movements and temperature variations. The curvature of the arch means that thick preformed plates will be difficult to apply, and thinner preformed plates or *in-situ* strengthening methods are more appropriate. The peel stresses resulting from curvature of the fibres require careful evaluation.

Jack-arch construction

In jack-arch floor and bridge decks, the dead to live load ratio is relatively high. For the FRP to have maximum effectiveness, it is desirable to reduce the permanent stress in the cast iron by load-relief jacking of the deck before applying the FRP strengthening material. It is very difficult to strengthen the beam for shear because the jack arches obstruct access. Should shear strengthening be necessary, then the jack arches could be removed to provide access and be replaced by a more modern form of decking, although this is unlikely to be economic.

Connections

The most difficult areas to strengthen in a cast iron structure are the connections, where there are stress concentrations and the stress field is often three-dimensional. Prestressed FRP is not viable over short lengths, and the adhesive bond between the FRP and the substrate is likely to be more critical, due to the high stress gradient in the vicinity of the connections. Access to the critical surfaces is also often restricted or impossible.

Where the load capacity of a connection detail needs to be increased, it is desirable to assume that the cast iron can crack and that the FRP absorbs the released stress resultants (see Section 5.2.5). Due to the relatively small extent of the stress concentration zone, the energy released by cracking at the point of maximum stress can be quite low compared with the energy released by a beam section when it cracks. The strengthening material can then be designed such that it does not debond if the cast iron fails, and hence maintains the structural integrity of the connection. The reduction in stiffness of the connection after the cast iron cracks should be considered.

FRP material should be added to provide load path continuity through the connection detail. Wet lay-up FRP can be more easily tailored to suit the connection detail, although FRP plates may also be used. The surface to be strengthened should first be regularised using filler adhesive to provide a flat surface, with no abrupt changes in direction.

4.4.2 Steel and ductile cast iron structures

Carbon steel and ductile cast iron are able to undergo a considerable amount of plastic flow as they yield. Therefore, a lower-modulus FRP material can be used to strengthen steel, since the strain and stress capacity of the FRP can be mobilised after the substrate has yielded, and the strengthening material can be used considerably more effectively than in cast iron.

The absence of substrate cracking means that the issue of debonding at cracks is of no consequence and delamination is only likely at the ends or edges of the strengthening plate, or at sections of high shear flow. Ultimate failure is likely to occur first in the FRP or in the adhesive bond, although the steel is likely to reach yield well before then.

To avoid permanent deformation of the steel, it will normally be appropriate to dimension the strengthening material to avoid yield of the steel under serviceability loading, as well as to satisfy safety with respect to fibre rupture and plate debonding under ultimate load conditions.

Steel structures are generally composed of thin-walled sections that present local and global buckling problems. The strength of non-compact steel members is frequently dictated by buckling, which in turn is a function of the stiffness of the sectional elements (see Section 4.1.1).

4.4.3 Wrought iron structures

Wrought iron is a ductile material, typically with similar yield stress and ultimate strain to those of mild steel. The conceptual design will therefore be similar to that for steel, although the possibility of delamination failure within the wrought iron must be considered (Section 2.2.3).

4.5 BIBLIOGRAPHY

Bassetti, A, Nussbaumer, A and Hirt, M (2000). "Crack repair and fatigue life extension of riveted bridge members using composite materials". In: A-H Hosny (ed), *Proc bridge engg conf 2000, ESE-IABSE-FIB, Sharm El-Sheikh*. Egyptian Soc Engrs, Cairo, vol I, pp 227–238

Concrete Society (2000). *Design guidance for strengthening concrete structures using fibre composite materials*. Technical Report 55, Concrete Society, Crowthorne

Moy, S S J, ed (2001b). *FRP composites – life extension and strengthening of metallic structures*. ICE design and practice guide, Thomas Telford, London

5 Structural behaviour and analysis

> This chapter describes the structural behaviour of a metallic structure strengthened with FRP, identifying basic assumptions and potential modes of failure. It presents the analysis methods needed for predicting structural behaviour, an essential step in the design of FRP strengthening for metallic structures. These include sectional analysis, covered in Section 5.2, and analysis of the adhesive joint, covered in Section 5.3. Two alternative approaches to adhesive joint analysis are presented: a stress-based approach and a fracture mechanics-based approach. Consideration is given to the cases of brittle and ductile metallic structures and non-prestressed and prestressed strengthening.
>
> Sizing methodologies are covered in this chapter; the factors of safety that dictate quantities of strengthening required are covered in Chapter 6. Detailed step-by-step design procedures are provided for the guidance of the designer. Derivations of the formulas presented are given in Appendices 5, 6 and 7, and a worked example is included in Appendix 9.

The nomenclature used in this chapter can be found on pages 15-17. "f" refers to the FRP material; "s" refers to the metallic substrate.

5.1 BEHAVIOUR OF METALLIC STRUCTURES STRENGTHENED WITH FRP MATERIALS

The behaviour of an FRP-strengthened metallic structure is described by the following principal characteristics.

1. Differential thermal expansion effects must be considered. There can be a significant difference between the coefficients of thermal expansion of the composite material and the metal, resulting in substantial restrained thermal stresses.

2. Effective strengthening requires an FRP material of adequate strength and stiffness in relation to the metal.

3. The adhesive joint between the FRP and the metallic substrate is often the weakest link and may control the mode of failure.

4. Where multi-material FRP is used, eg carbon FRP containing a glass FRP layer (Section 6.6.7), the properties of the combined composite should be considered in the structural analysis.

5. Metallic structures are often susceptible to buckling.

6. The permanent stresses in the structure at the time of application of the strengthening must be considered in the analysis. This is particularly important for brittle materials such as cast iron.

The above characteristics mean that the behaviour of a metallic structure strengthened with FRP differs significantly from that of concrete structures strengthened with FRP.

The sensitivity of a design to uncertainties in the properties of the existing structure should be assessed. These include uncertainty in material properties, structural geometry, connections between structural elements, restraints, permanent stresses and temperature. If it is not possible to predict accurately the permanent state of stress in a structure by analysis due to the uncertainties set out above, a structural investigation programme should be undertaken to determine the necessary data.

5.1.1 Potential failure modes

Potential failure modes of an FRP-strengthened structure include:

(a) tensile rupture of the strengthening material

(b) tensile rupture or yielding of the metal substrate

(c) adhesion failure in the adhesive joint, substrate, or within the FRP, at a point of high shear stress such as the end or a curtailment point of a strengthening element

(d) adhesion failure at a crack or other discontinuity in the substrate

(e) global buckling of the strengthened member

(f) local buckling of the strengthened member

(g) compressive failure of the strengthening material

(h) fatigue failure in the metal, the FRP, or the adhesive interface

(i) creep rupture failure of the FRP or the adhesive interface

(j) failure due to loss of prestress, resulting from creep of the adhesive

(k) accidental damage of the FRP and substrate due to impact or fire.

Failure modes (a) to (h) are illustrated in Figure 5.1. Modes (i) and (j) describe failure of the strengthened member due to creep of the strengthening elements under sustained long-term loading. Accidental damage (k) results in some loss of the FRP material. Note that failure modes (h) to (k) are critical only if they lead to failure of the metallic member. In many cases this can be avoided if remedial maintenance is carried out.

5.2 SECTIONAL ANALYSIS OF A STRENGTHENED MEMBER

Sectional analysis allows longitudinal stresses and strains in the strengthening material and the substrate to be determined, assuming that the FRP and metal act compositely. These stresses, in conjunction with limiting values appropriate for the limit state under consideration, allow the quantity of FRP material required to strengthen a beam or column structure to be calculated.

Loads should be factored by the appropriate load partial factor, and the material properties by the appropriate material partial factor. This is covered in Section 6.1.

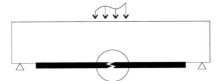
(a) Tensile rupture of strengthening

(e) Global member buckling

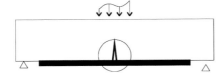
(b) Tensile rupture of substrate

(f) Local buckling of member

Buckling of web

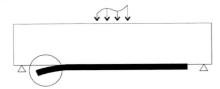
(c) Delamination of ends of strengthening

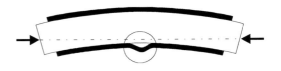
(g) Compressive failure of laminate

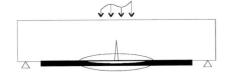
(d) Delamination at a substrate discontinuity

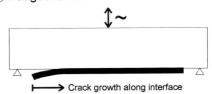
(h) Fatigue failure

Crack growth along interface

Figure 5.1 *Potential failure modes for flexural FRP strengthening*

5.2.1   **Permanent stress at time of application of strengthening**

As discussed in Section 4.2, it is necessary to take into account in the analysis the stress in the existing member at the time of application of the strengthening material. The stress includes the effects of:

- permanent loads
- temperature
- support settlements
- temporary works loads.

There are two possible approaches to the sectional analysis, which take into account the permanent stress in the existing member at the time of strengthening: (a) the initial stress method, and (b) the initial strain method.

Initial stress method

The *initial stress method* considers the additional load applied to the strengthened structure after the FRP has been applied. The stresses due to this additional load are superimposed on the initial stresses present in the structure before the FRP was applied (Figure 5.2).

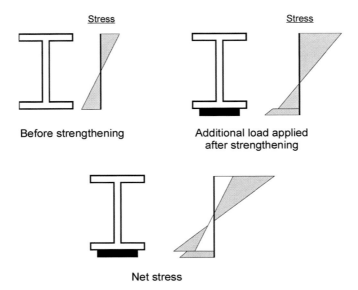

Before strengthening

Additional load applied after strengthening

Net stress

Figure 5.2 *The initial stress method*

The uniaxial constitutive laws for the initial stress method are as follows.

For the metallic substrate:

$$\sigma_s = \sigma_{s0} + E_s\left(\Delta\varepsilon_s - \alpha_s\Delta T_s\right) \text{ for elastic behaviour} \tag{5.1}$$

or

$$\sigma_s = fn\left(\frac{\sigma_{s0}}{E_s} + \Delta\varepsilon_s - \alpha_s\Delta T_s\right) \text{ for non-linear, or elasto-plastic, behaviour} \tag{5.2}$$

(where *fn* is a non-linear function, describing the stress-strain curve of the metal).

For the FRP:

$$\sigma_f = E_f\left(\Delta\varepsilon_f - \alpha_f\Delta T_f\right) \tag{5.3}$$

where:

σ_{s0} is the stress in the substrate at the time of strengthening

E is the Young's modulus

$\Delta\varepsilon$ is the incremental strain

α is the coefficient of thermal expansion

ΔT is the change in temperature.

Initial strain method

The *initial strain method* considers the total load applied to the composite section, including both the loads at the time of strengthening and the additional load applied since strengthening. The difference in strain between the substrate and the FRP when the strengthening is applied is defined as ε_{f0}, as illustrated in Figure 5.3. ε_{f0} includes the effect of any prestress applied to the FRP.

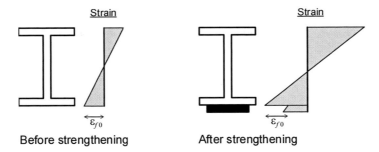

Figure 5.3   *The initial strain method*

After strengthening, the difference in strain between the substrate and the FRP composite will always be ε_{f0}, so that

$$\varepsilon_f = \varepsilon_s - \varepsilon_{f0} \tag{5.4}$$

where ε_s is the total strain in the substrate at the level of the FRP.

The uniaxial stress-strain laws for the initial strain method are, for the metallic substrate:

$$\sigma_s = E_s\left(\varepsilon_s - \alpha_s T_s\right) \text{ for elastic behaviour} \tag{5.5}$$

or:

$$\sigma_s = fn(\varepsilon_s - \alpha_s T_s) \text{ for non-linear, or elasto-plastic behaviour} \tag{5.6}$$

(where *fn* is a non-linear function, describing the stress-strain curve of the metal), and, for the FRP:

$$\sigma_f = E_f\left(\varepsilon_f - \alpha_f T_f\right) \tag{5.7}$$

The initial strain method is used in this report, as it is generally more straightforward to apply.

5.2.2   Material characteristics

The effects of differential thermal expansion should be taken into account in the sectional analysis (see Section 6.3.3), as there can be a significant difference in the coefficients of thermal expansion of the metal and the FRP. For this reason, it is important to consider the temperature terms in the material constitutive laws, given by Equations 5.1–5.7.

The capacity of the strengthened section is governed by the most critical of the limiting strains permitted for the metal substrate and the FRP. These are:

1.   for the substrate:

$$\bar{\varepsilon}_{sc} \leq \varepsilon_s \leq \bar{\varepsilon}_{st} \tag{5.8}$$

2.   for the FRP:

$$\bar{\varepsilon}_{fc} \leq \varepsilon_f \leq \bar{\varepsilon}_{ft} \tag{5.9}$$

with subscripts "c" and "t" referring to the limiting compressive and tensile strains respectively. The overbar is used to indicate a limiting value. $\bar{\varepsilon}$ is a limiting strain with reference to the ultimate or serviceability limit state as appropriate. It is determined from the ultimate strain to failure, ε_u, and the material partial factor (Section 6.1.4), γ_m, appropriate to the limit state, by

$$\bar{\varepsilon} = \frac{\varepsilon_u}{\gamma_m} \qquad (5.10)$$

The compressive and tensile ultimate strengths of an FRP strengthening product should be provided by the materials supplier, or obtained from tests (Section 7.1.1). The compressive failure strain of the laminate can be much lower than its tensile failure strain.

5.2.3 Sizing procedure for FRP strengthening of a brittle elastic beam

For an existing, unstrengthened member made from a brittle material (such as cast iron), the critical case is likely to be the initiation of a tensile crack in the tension face of the member. If the FRP is prestressed, however, failure could also occur by tensile rupture of the FRP. The procedure outlined in Figure 5.4 shows the stages required to size the cross-sectional area of FRP material needed to increase the bending capacity of a member to M_u.

The procedure is based on the assumptions that:

- there is full composite action between FRP and metallic member
- the stiffness of the adhesive layer is negligible
- plane sections remain plane
- the stress-strain behaviour of the metal member is linear-elastic
- the FRP plate is relatively thin compared with the member being strengthened.

The sectional analysis given here assumes that there is no variation in strain across depth of the FRP plate. If a relatively thick FRP plate is used, the analysis must be adapted by include the flexural stiffness of the FRP, EI_f (see the figure in Step 4, below).

All moments in Section 5.2.3 are defined as positive in sagging (as shown in the figures below).

A worked example is included in Appendix 9, which includes the sectional design of a cast iron beam strengthened using CFRP using the method presented in this section.

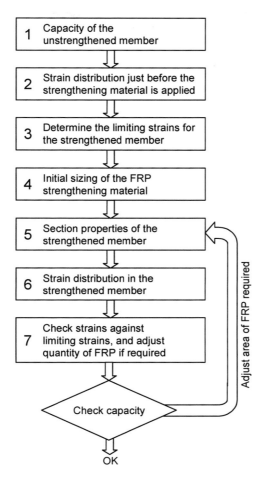

Figure 5.4   *The sectional design procedure*

Step 1 – Determine the capacity of the unstrengthened member

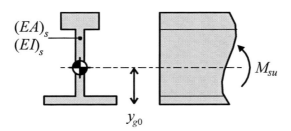

The moment capacity of the unstrengthened member will have been determined during assessment of the structure (when it was determined that strengthening is required). It is given by:

$$M_{su} = \frac{(EI)_s \bar{\varepsilon}_{st}}{y_{g0}}$$

(5.11)

$\bar{\varepsilon}_{st} = \sigma_{tu} / E_s$ is the factored limiting tensile strain of the metal (which can be found from tests or appropriate guidance, such as Highways Agency, 2001a).

Step 2 – Determine the state of stress and strain in the unstrengthened member at the time of application of strengthening

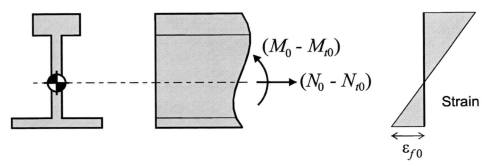

At the time of application of the FRP, the stress resultants acting on the member can be decomposed into two components:

(N_0, M_0) Permanent loads acting on the member at the time of strengthening. (N_0, M_0) also include the stress resultants in the metallic member due to prestress in the FRP, and the statically indeterminate effects of thermal loading.

(N_{t0}, M_{t0}) Stress resultants due to the fully restrained temperature effects. These result from the difference between the temperature at the time of strengthening, T_0 and the temperature at the time the metallic member was installed, T_i (see Section 6.2.1).

N_{t0} and M_{t0} are given by:

$$N_{t0} = -E_s \alpha_s \int_{member} (T_0 - T_i)\, b.\, \mathrm{d}y \tag{5.12}$$

$$M_{t0} = -E_s \alpha_s \int_{member} (T_0 - T_i)\, b\, y.\, \mathrm{d}y \tag{5.13}$$

where y is measured from the centroid of the section. In general T_0 is a function of y, however, for a uniform change in temperature of $(T_0\text{-}T_i)$ throughout the member:

$$N_{t0} = -(EA)_s \alpha_s (T_0 - T_i) \tag{5.14}$$

$$M_{t0} = 0 \tag{5.15}$$

The strain at the position where the strengthening material is to be applied (ie the tension face) is

$$\varepsilon_{s0} = \frac{N_0 - N_{t0}}{(EA)_s} + \frac{M_0 - M_{t0}}{(EI)_s} y_{g0} \tag{5.16}$$

and the stress is:

$$\sigma_{s0} = E_s \left\{ \frac{N_0 - N_{t0}}{(EA)_s} + \frac{M_0 - M_{t0}}{(EI)_s} y_{g0} - \alpha_s (T_0 - T_i) \right\} \tag{5.17}$$

Step 3 – Establish limiting strains in the strengthened member

Tensile failure can occur at the tension face in either the metallic member or the FRP plate. The limiting strain at the tension face of the metal member is thus:

$$\bar{\varepsilon}_{yg0} = \min\left(\bar{\varepsilon}_{st}, \ \bar{\varepsilon}_{ft} + \varepsilon_{f0}\right) \tag{5.18}$$

The difference in strain between the FRP and substrate, ε_{f0}, is given by Equation 5.4, in which $\varepsilon_s = \varepsilon_{s0}$ and $\varepsilon_f = \varepsilon_{fp}$ describes any prestrain in the FRP:

$$\varepsilon_{f0} = \varepsilon_{s0} - \varepsilon_{fp} \tag{5.19}$$

Step 4 – Initial sizing of the FRP strengthening material

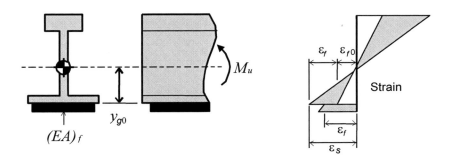

For the case of a predominantly flexural member, the required area of FRP strengthening material is **estimated** by:

$$A_f = \frac{E_s}{E_f} \left(\frac{M_u - M_{su}}{M_{su} - M_0}\right) \left(\frac{1}{1 + \left(y_{g0}^2 A_s / I_s\right)}\right) A_s \tag{5.20}$$

where:

M_u is the required moment capacity

M_{su} is the moment capacity of the unstrengthened member

M_0 is the moment at the time the strengthening becomes active.

This equation is based on bending only, and assumes that the capacity is limited by the strain in the metallic member adjacent to the adhesive joint. As a consequence, the equation cannot be used with prestressed strengthening, which applies a compressive axial load to the metallic member. Differential thermal expansion effects and axial loads are taken into account in the subsequent check analysis. In practice, it will generally be more convenient to use a spreadsheet to size the amount of FRP required, in which the area of FRP is adjusted until the strain limits (Step 7) are satisfied.

Step 5 – Section properties of the strengthened member

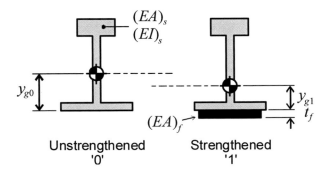

Unstrengthened
'0'

Strengthened
'1'

The section stiffness properties of the strengthened member are:

$$(EA)_1 = (EA)_s + (EA)_f \tag{5.21}$$

$$y_{g1} = \frac{y_{g0}(EA)_s - 0.5t_f(EA)_f}{(EA)_1} \tag{5.22}$$

$$(EI)_1 = (EI)_s + (EA)_s(y_{g0} - y_{g1})^2 + (EA)_f(y_{g1} + 0.5t_f)^2 \tag{5.23}$$

Step 6 – Calculate strains in the strengthened member

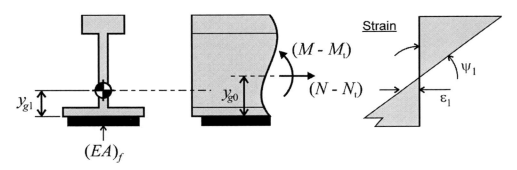

The stress resultants that the strengthened member is required to carry can be decomposed into two components (in a similar manner to the unstrengthened case, described in Step 2).

(N, M) the externally applied loads (including both live and dead loads) and the statically indeterminate effects of thermal loading

(N_t, M_t) The stress resultants due to the fully restrained temperature effects, at the operating temperature, T (see Section 6.2.1). These are in part due to differential thermal expansion between the FRP and the substrate.

These resultants are referred to the centroid of the *unstrengthened* section (y_{g0}).

$$N_t = -E_s\alpha_s \int_{member}(T - T_i)\,b\ \mathrm{d}\,y + (EA)_f\alpha_f(T - T_0) \tag{5.24}$$

$$M_t = -E_s\alpha_s \int_{member}(T - T_i)\,b\,y\ \mathrm{d}\,y + (EA)_f\alpha_f(T - T_0)(y_{g0} + 0.5t_f) \tag{5.25}$$

The reference temperature in the FRP is T_0, the temperature at the time of strengthening. In the metallic member, the reference temperature is T_i, the temperature at the time of installation.

For a uniform temperature in the metallic member of T_s, and a temperature of T_f in the FRP:

$$N_t = -(EA)_s \alpha_s (T_s - T_i) - (EA)_f \alpha_f (T_f - T_{f0})$$ (5.26)

$$M_t = -(EA)_f \alpha_f (T_f - T_{f0})(y_{g0} + 0.5t_f)$$ (5.27)

The strain distribution in the strengthened member is defined by the strain at the level of the centroid of the composite member, ε_1, and the curvature, ψ_1:

$$\varepsilon_1 = \frac{N - N_t + (EA)_f \varepsilon_{f0}}{(EA)_1}$$ (5.28)

$$\psi_1 = \frac{(M - M_t) - (N - N_t)(y_{g0} - y_{g1}) + (EA)_f \varepsilon_{f0}(y_{g1} + 0.5t_f)}{(EI)_1}$$ (5.29)

Step 7 – Check strains against limiting strains

Equation 5.20 gives an *estimate* of the area of FRP required to strengthen a member. However, it does not consider differential temperature effects or failure in the metallic member (rather than the FRP). The extreme strains calculated in Step 6 must be compared with the allowable limiting strains in both metallic substrate and the FRP strengthening material:

$$\varepsilon_1 + \psi_1 y_{g1} - \alpha_s (T_s - T_i) \le \bar{\varepsilon}_{st}$$ (5.30)

$$\varepsilon_1 + \psi_1 y_{g1} - \alpha_f (T_f - T_{f0}) - \varepsilon_{f0} \le \bar{\varepsilon}_{ft}$$ (5.31)

Limiting strains for the metallic substrate can be found, for example, in BD 21/01 (Highways Agency, 2001a).

If the limit strains are satisfied, the strengthened member will have the required moment capacity, M_u. Otherwise, the area A_f must be adjusted and the process repeated from Step 5 until the required moment can be carried.

The sectional design process shown in Figure 5.4 is iterative. However, this is not a major problem, as the formulae can easily be entered into a spreadsheet. The area of strengthening can then easily be adjusted, either manually or by using the spreadsheet's solver function.

5.2.4 Sizing FRP strengthening for increased stiffness with a brittle elastic beam

The area of FRP required to increase the axial and flexural stiffness of a member can be assessed using Equations 5.21 to 5.23 in Step 5.

Cast-iron section, taking cracking into account

The approach used in Section 5.2.3 limits the tensile strain in the brittle metal such that cracking does not occur. This is in contrast to the approach used to design a reinforced concrete member strengthened with FRP, in which the substrate is allowed to crack, due to the presence of internal reinforcement (as shown in Figure 5.5). However, in a few cases it may be appropriate to use a cracked-section analysis. It should be noted that some major owners of structures, such as Network Rail, do not allow cracking of the cast iron under any circumstances.

A cracked-section analysis should not generally be applied to primary members, in which cracking is catastrophic (for example, tests by Moy *et al*, 2000). For example, a cracked-section analysis must not be used for primary I-section beams in a bridge.

For FRP to strengthen a cracked section successfully the tensile actions carried by the metal before the crack forms must be transferred to the FRP after the crack forms, *without delamination*. The adhesive connections must be able to absorb the energy released as the metal member cracks.

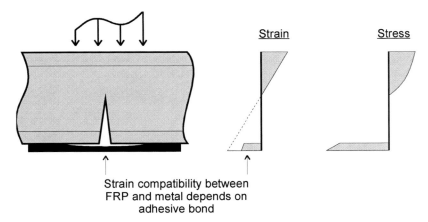

Strain

Stress

Strain compatibility between
FRP and metal depends on
adhesive bond

Figure 5.5   *Sectional analysis of a cast iron strengthened section, which is allowed to crack*

In the case of smaller members less energy is released during crack formation, with the result that when the metallic member cracks, the FRP may not debond explosively but remain integral with the cracked section in the same way as it would with reinforced concrete. This is supported by some tests that have been carried out on strengthened sections with existing cracks in which the FRP retained its integrity as in an uncracked section.

If the cracked section is stable, the strengthened section may be designed on the basis of the FRP acting compositely with the cracked section. For example, a cracked-section analysis was applied to the secondary spandrel ring members of Tickford Bridge (Appendix 1).

Sectional analysis of a cracked section should be carried out using a non-linear stress-strain curve for cast-iron (Figure 2.5). The capacity in this case is governed by either the limiting tensile strain in the FRP or the limiting compressive strain in the cast iron. The adhesive joint analysis described in Section 5.3 must be used to assess the stability of the adhesive interface for a cracked section.

Note that a small amount of delamination of the FRP from the substrate is necessary in the vicinity of the crack, to avoid high strains in the FRP (unless the displacement associated with crack opening can be fully accommodated by shear deformation of the adhesive layer). The stability of the cracked section for different lengths of delaminated FRP should be checked in the adhesive joint analysis.

If the member to be strengthened is critical to the global stability of the structure, analysis of the section should be backed up by physical tests. These tests should characterise delamination of the FRP from the substrate being used. A minimum of five tests is recommended.

5.2.6   Sizing procedure for FRP strengthening of an elasto-plastic beam

Steel, wrought iron and ductile cast iron are ductile materials. Considerable plastic strain can develop before failure of the metallic section occurs. Where a compact metallic member is being strengthened, an elasto-plastic design method can be used. It is assumed that the metallic section behaves elasto-plastically, while the FRP strengthening behaves elastically.

Failure of the strengthened member will usually occur by tensile rupture of the FRP before the ultimate tensile rupture strain of the steel is reached. Alternatively, delamination could occur within the adhesive joint due to high shear/peel stresses on the interface. Failure could also occur by local buckling of the compression flange or web, or by global, flexural-torsional buckling (in cases where the compression flange of the beam is not adequately restrained).

Research into strengthening metallic structures using FRP has to date focused on strengthening brittle cast iron. Experimental work on strengthening ductile steel members is reported in Sen *et al* (1995, 2001), Liu, Silva and Nanni (2001), Mertz *et al* (2001) and Miller *et al* (2001).

A method for designing FRP strengthening for an elasto-plastic design of a beam has not been generally agreed or validated. However, sectional analysis of a strengthened, elasto-plastic beam generally follows the same philosophy as for a brittle beam. Steps 1–4 from Section 5.2.3 can be used to estimate the FRP cross-sectional area required (but now using the plastic moment capacity of the ductile metal section in Equation 5.11).

Figure 5.6 shows the strain and stress distribution in a strengthened beam for a substrate exhibiting perfectly plastic behaviour. Some substrate materials exhibit post yield strain hardening, though the design curves specified in structural codes frequently neglect this and assume perfectly plastic behaviour.

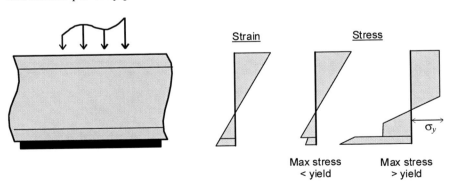

Figure 5.6   *Sectional analysis of a strengthened steel section*

A failure envelope in terms of N versus M can be constructed for the strengthened section for a given cross-sectional area of FRP, using:

$$\overline{N} = EA_f\left(\varepsilon_1 + \psi_1\, y_f - \alpha_f\, \Delta T_f - \varepsilon_{f0}\right) + \int_A \sigma_s\left(\varepsilon_1 + \psi_1\, y - \alpha_s\, \Delta T\right) dA \qquad (5.32)$$

$$\overline{M} = EA_f\left(\varepsilon_1 + \psi_1\, y_f - \alpha_f\, \Delta T_f - \varepsilon_{f0}\right) y_{g0} + \int_A \sigma_s\left(\varepsilon_1 + \psi_1\, y - \alpha_s\, \Delta T\right) y\, dA \qquad (5.33)$$

The integrals in these equations extend over the whole member section. It is important to note that the strain in the FRP is reduced by the subtraction of the strain in the substrate at the time of application, ε_{f0}.

Each point on the failure envelope corresponds to a distinct position of the neutral axis and a corresponding pair of strain parameters ε_1 and ψ_1 that satisfy the most critical limiting strain constraint.

If the applied stress resultants (N, M) lie outside the failure envelope $(\overline{N}, \overline{M})$, the cross-sectional area of FRP A_f will need to be increased.

5.2.7   Sectional analysis of a strengthened steel-concrete composite section

A steel-concrete composite section can also be examined using an elasto-plastic analysis. The analysis follows the same philosophy as described for elasto-plastic sections, with the composite concrete element also being taken into account in the sectional analysis. The concrete and steel are both assumed to be elasto-plastic, whilst the FRP remains elastic. Typical stress and strain distributions in a strengthened section are shown in Figure 5.7.

Failure occurs either when the ultimate strain of the concrete is reached in compression, the FRP ruptures in tension, or the FRP debonds from the metallic section. As with the strengthening of concrete structures using FRP (Concrete Society, 2000), a balanced section is defined as one in which failure of the concrete and failure of the FRP are reached at the same time. When the design moment is above the balanced section moment, failure will occur first in the concrete; below the balanced moment, failure will occur by rupture of the FRP.

The strengthening of steel-concrete composite sections using bonded FRP has been investigated experimentally by Sen *et al* (1995, 2001). A ductile failure, accompanied by considerable deflection, was obtained. It was found necessary to include mechanical anchorages between the FRP and the steel beam to avoid brittle, unstable, adhesion failure. These tests illustrate the importance of checking that the adhesive bond does not fail, the procedure for which is given in Section 5.3.

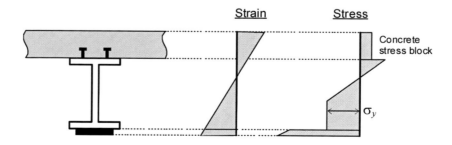

Figure 5.7   *Sectional analysis of a steel-concrete composite strengthened section*

The buckling capacity of non-compact metallic members strengthened using externally bonded FRP can be assessed using standard buckling techniques and formulae on a composite section basis. The stiffness properties of the strengthened section (for example, the flexural stiffness, EI, and torsional stiffness, GJ) should be calculated by summing the contributions of the metallic member and the FRP (as in Section 5.2.3). It should also be checked that the FRP does not fail in compression before the elastic buckling capacity of the strengthened member is reached. Moy (2001b) gives formulae for some specific buckling modes.

For example, the elastic flexural-torsional buckling resistance of a beam is given by:

$$M_{cr} = \frac{\pi}{L} \sqrt{(EI)_{yy} \left(GJ + (EI)_{ww} \frac{\pi^2}{L^2} \right)}$$

(5.34)

where $(EI)_{yy}$, GJ, and $(EI)_{ww}$ are the lateral, torsional and warping rigidities of the section. These should be calculated taking into account the contribution of the FRP with its appropriate modulus. An I-section beam having a second moment of area about its vertical axis of I_{ys} strengthened by a FRP plate on the bottom flange having a second moment of area of I_{yf} has a composite $(EI)_{yy}$ value of:

$$(EI)_{yy} = E_s I_{ys} + E_f I_{yf}$$

(5.35)

The local buckling strength of a simply supported rectangular plate is given by:

$$N_{x,cr} = 2 \left(\frac{\pi}{b} \right)^2 \left(\sqrt{D_{11} D_{22}} + D_{12} + 2 D_{66} \right)$$

(5.36)

where D_{11}, D_{22}, D_{12}, and D_{66} are the flexural rigidities of the plate and b is the transverse width of the plate. In the case of a metallic plate of thickness t_s having layers of FRP on each face of thickness t_f, the flexural stiffness terms of the plate are given by expressions of the form:

$$D_{11} = \frac{E_s t_s^3}{12 \left(1 - v_s^2 \right)} + \frac{1}{2} \left(\frac{E_{xf}}{1 - v_{xyf} v_{yxf}} \right) t_f \left(t_s + t_f \right)^2$$

(5.37)

$$D_{22} = \frac{E_s t_s^3}{12 \left(1 - v_s^2 \right)} + \frac{1}{2} \left(\frac{E_{yf}}{1 - v_{xyf} v_{yxf}} \right) t_f \left(t_s + t_f \right)^2$$

(5.38)

5.3 ADHESIVE JOINT ANALYSIS

The sectional analysis (Section 5.2) considers only normal longitudinal stresses and assumes that the FRP and metal act fully compositely at the section under consideration. However, shear and peel stresses also develop in the adhesive layer, and these need to be considered separately. Local shear and peel stress concentrations develop where there is a discontinuity in the substrate, the adhesive or the FRP strengthening, as illustrated in Figure 5.8. The bond between the FRP and the metallic substrate is often the weakest element in a strengthening scheme (as shown by the experimental work of Liu *et al* (2001) and Sen *et al* (2001)).

Differential thermal expansion between the metallic substrate and the FRP can lead to high adhesive stresses, and these may dominate the design of the adhesive joint. It is very important therefore to consider temperature effects during the design of externally bonded FRP strengthening (see Section 6.3.1).

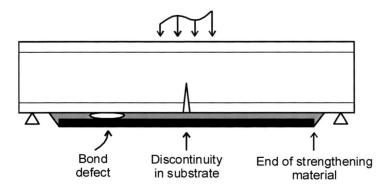

Bond defect

Discontinuity in substrate

End of strengthening material

Figure 5.8   *Strain discontinuities in the substrate, adhesive and strengthening material*

At the time of writing there is no independent guidance for analysing the adhesive joint in a metallic structure strengthened using FRP materials in engineering practice. However, relevant models have been in use for several decades to determine the bond stresses acting between dissimilar materials, in civil and other engineering applications. Examples include bond stresses between concrete and steel reinforcing bars in reinforced concrete, and adhesive joints in aerospace structures.

Two approaches can be used to assess the strength of the adhesive layer:

- an elastic, stress-based, analysis (Section 5.3.1)
- a fracture mechanics, energy-based approach (Section 5.3.2).

These two approaches are compared in Table 5.1.

Table 5.1   *Comparison of the elastic and fracture mechanics methods for examining the adhesive joint*

	Elastic, stress-based analysis	Fracture mechanics, energy-based analysis
Characteristics	• Elastic analysis • The elastic shear and normal stress distribution is determined along the adhesive joint • The thickness and shear modulus of the adhesive are required	• Ultimate analysis • The energy release rate $\overline{G}_{IC}$ at failure of the adhesive joint is calculated • The adhesive thickness and shear modulus are not required
Advantages	• Can be applied to FRP plates of varying thickness	• The detailed stress distribution does not need to be determined • Formulae easier to use
Disadvantages	• Elasto-plastic behaviour of adhesive not taken into account • Thickness and shear modulus of the adhesive layer are not accurately known • Results are sensitive to bond defects	• Fracture mechanics is not widely used in civil engineering • More development required for beams of non-constant cross-section • Does not fully address temperature effects

The fracture mechanics method leads to a simple design formula that characterises the energy release during failure of a bonded joint, without needing to consider the stress distribution along the joint. The simplicity of this approach makes it of benefit in design, but it is limited to a beam and plate of constant cross-section and needs to be developed further if it is fully to take temperature effects into account.

The stress-based method calculates the stress distribution in the adhesive joint. It can be applied to complex strengthening arrangements, such as beams with varying cross-section or FRP plates that taper along the beam. A more complex analysis of the adhesive joint is required than for the fracture mechanics approach. The results of lap-shear tests (used to characterise the shear strength of a bonded joint) require back-analysis before they can be used in design, but this can be implemented easily in a spreadsheet or incorporated into finite element models for more complex strengthening schemes.

5.3.1 Elastic, stress-based analysis

An elastic analysis based on an elastic characterisation of the adhesive layer is used to determine the distribution of stress along the adhesive joint. Figure 5.9 shows predicted stress distributions near the end of a strengthening plate from such an analysis for a particular case. The most significant stresses in the adhesive layer are the longitudinal shear stress, τ, and the through thickness normal stress, σ. A tensile normal stress may lead to peeling of the FRP away from the substrate, and is also referred to as *peel stress*.

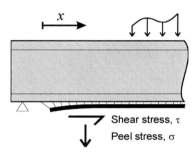

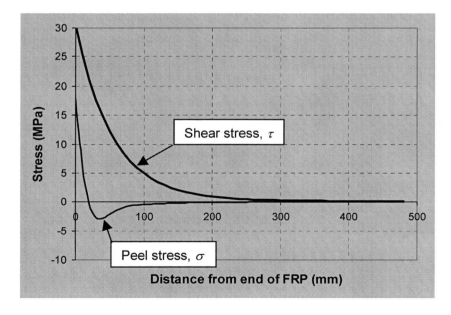

Figure 5.9   *Example of shear and peel stress distributions in the adhesive near the end of a strengthening plate (produced from the worked example, in Appendix 9)*

In the example shown, the peel stress is negligible compared with the shear stress, due to the use of an FRP plate of small thickness. This might not be the case were a thicker plate to be used.

Failure of the adhesive joint will initiate at the position of maximum principal stress. In Figure 5.9, this occurs at the end of the FRP plate. The strength of the joint is normally dictated by the strength of the adhesive, or the interlaminar shear strength of the FRP, since these are usually lower than the shear/interlaminar shear strength of the metallic substrate, unless the latter has been degraded by corrosion. In the presence of combined shear and peel stresses, failure of the adhesive layer can be characterised by the maximum principal stress σ_1:

$$\sigma_1 = \frac{\sigma}{2} + \sqrt{\left(\frac{\sigma}{2}\right)^2 + \tau^2} \leq \bar{\sigma} \tag{5.38}$$

$\bar{\sigma}$ is the characteristic strength of the adhesive system, including any primer, obtained by back-analysis of lap-shear test results (described below). In design, σ_1 should incorporate partial load factors and $\bar{\sigma}$ incorporates material partial factors appropriate to the limit state under consideration.

Figure 5.10 outlines the procedure for designing an adhesive joint, the details of which follow. The equations given here are only valid for a strengthened member of constant cross-section, but the procedure can be applied to any adhesive joint.

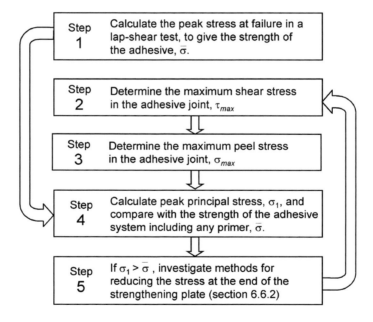

Figure 5.10 *Analysis of an adhesive joint, using an elastic, stress-based method*

The equations given below are taken from Appendix 5. They have been checked as far as possible by a desk study, but experimental work has not been carried out to validate the analysis.

Although the equations are of closed-form, they are best evaluated by spreadsheet for ease of use in design. The loads should be factored by the appropriate load partial factor, and the material properties by the appropriate material partial factor, as covered in Section 6.1.

The worked example in Appendix 9 includes an adhesive joint analysis using the method given below.

Step 1 – Determine the strength of the adhesive system from the lap-shear test result

A lap-shear test (described in Section 7.1.1) gives the value of the failure load of the sample, P, which is often expressed as an *average* shear stress τ_{avr} over the entire bonded area. However, it is the peak stress that causes bond failure (as shown in Figure 5.11). This peak stress can be calculated by back-analysis of the lap-shear results, using the equations in Appendix 6. This approach allows the adhesive strength to be determined for any lap-shear test arrangement, as the geometry and properties of the substrate are explicitly taken into account in the analysis.

It is important that the lap-shear test coupon is representative of the adhesive system in the works. It should therefore include any primer layer.

It is important to appreciate that it is not meaningful to compare the peak shear or principal stress determined from the adhesive joint analysis with the average shear stress at failure obtained from a lap-shear test. The peak shear stress in an adhesive joint is usually considerably greater than the average shear stress.

From the lap-shear test result, determine the maximum shear stress, τ_{max}, and peel stress, σ_{max}, in the adhesive at failure (from Appendix 6). Using these values of τ_{max} and σ_{max}, calculate the adhesive strength in terms of the maximum principal stress at failure $\overline{\sigma}$.

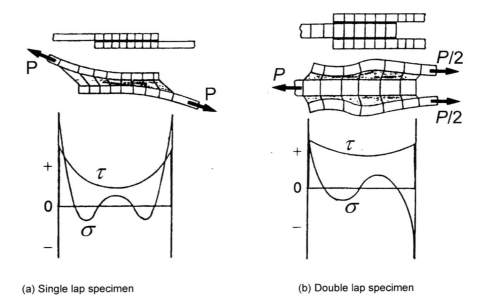

(a) Single lap specimen   (b) Double lap specimen

Figure 5.11 *Single lap and double lap-shear tests, showing the distribution of shear (τ) and peel (σ) stresses within a lap-shear specimen (from Mays and Hutchinson, 1992)*

Step 2 – Calculate the maximum adhesive shear stress

The maximum adhesive shear stress, τ_{max}, in the adhesive joint due to a geometric discontinuity is calculated using the equations derived in Appendix 5, as given below.

Lack-of-fit strain, $\Delta\varepsilon_{fs}$

The lack-of-fit strain, $\Delta\varepsilon_{fs}$, is the difference in strain that would exist between the FRP and the metallic member if the adhesive bond were not present. This includes the effects of:

- additional loads applied to the structure after the adhesive has cured

- differential thermal expansion after the adhesive has cured (which can be very significant)

- prestress in the FRP strengthening material.

The applied loads are:

$(T-T_0)$ the temperature change since the strengthening (since the adhesive cured)

N_s the axial force resulting from loads applied to the metallic member *after the adhesive cured* (tension positive)

M_s the bending moment resulting from loads applied to the metallic member *after the adhesive cured,* applied at the neutral axis level of the metallic section (hogging positive)

N_{f0} the initial force in the FRP at the time of strengthening, due to prestress (tension positive). In non-prestressed schemes N_{f0} is zero at the end of a plate, or the force in the plate based upon a cracked sectional analysis at the location of a crack in the substrate

M_{f0} the initial bending moment in the FRP (hogging positive). This is usually zero, but should be used if the supplied FRP strengthening is initially curved, or if the plate is prestressed and the prestress force is not concentric with the plate.

The load and strain terms may vary along the beam, as a function of beam co-ordinate x. However, it will usually be sufficiently accurate to consider the situation at the end of the FRP plate or adjacent to a crack in the substrate.

If the adhesive bond were released, the applied loads (N_s, M_s) would be carried by the metallic section alone, giving a strain (ε_{sR}) and curvature (ψ_{sR}) in the metallic substrate of:

$$\varepsilon_{sR} = \frac{N_s}{E_s A_s} - \psi_{sR} y_s + \alpha_s \left(T_s - T_{s0}\right) \tag{5.39}$$

$$\psi_{sR} = \frac{M_s}{E_s I_s} \tag{5.40}$$

After the adhesive has cured, the prestress is transferred to the FRP, and this prestress is only maintained by the adhesive bond. If the adhesive bond is released, the FRP would contract back to its original length. The strain (ε_{fR}) and curvature (ψ_{fR}) in the FRP in the absence of bond would thus be:

$$\varepsilon_{fR} = -\frac{N_{f0}}{E_f A_f} + \psi_{fR} y_f + \alpha_f \left(T_f - T_{f0}\right) \tag{5.41}$$

$$\psi_{fR} = -\frac{M_{f0}}{E_f I_f} \tag{5.42}$$

The lack-of-fit strain is given by the difference in strain across the adhesive joint, if no adhesive bond was present. The strain can vary with x (along the joint).

$$\Delta\varepsilon_{fs}(x) = \varepsilon_{fR} - \varepsilon_{sR} \tag{5.43}$$

Flexibility parameters

The flexibility of the adhesive layer, the FRP and the metallic member are described by the parameters f_1 and f_2.

The shear flexibility of the adhesive layer is:

$$f_1 = \left(\frac{t_a}{G_a b_a}\right) \tag{5.44}$$

The relative shear flexibility of the FRP and the metallic member is:

$$f_2 = \left(\frac{1}{E_f A_f} + \frac{1}{E_s A_s} + \frac{z(z - t_a)}{E_s I_s + E_f I_f}\right) \tag{5.45}$$

The relative flexibility of the beam, FRP and adhesive is described by:

$$\lambda = \sqrt{\frac{f_2}{f_1}} \tag{5.46}$$

t_a is the thickness of adhesive

b_a is the width of the FRP plate

G_a is the shear modulus of adhesive

E_f is the longitudinal modulus of the FRP

A_f is the cross-sectional area of the FRP

E_s is the elastic modulus of the metallic member

I_s is the second moment of area of the metallic member

A_s is the cross-sectional area of the metallic member

z is the lever arm between the centroid of the unstrengthened metallic member and the centroid of the FRP plate

Maximum adhesive shear stress

The maximum adhesive shear stress is calculated using Equation A5.23 in Appendix 5:

$$\tau_{max} = -\frac{\Delta\varepsilon_{fs}\big|_{x=0}}{b_a\sqrt{f_1 f_2}} - \frac{\lambda N_f\big|_{x=0}}{b_a} \tag{5.47}$$

where $N_f\big|_{x=0}$ is the axial force in the FRP plate at the position of the strain discontinuity.

τ_{max} occurs at the end of a plate or at a crack in the substrate or adhesive, assumed to be located at $x = 0$. It should be noted that the above equation is applicable only when both the member and the FRP plate have a constant cross-section. In other situations a numerical analysis is required.

Equation 5.47 only includes the shear stress concentration due to the geometric discontinuity. The shear stress due to flexure of the beam must be calculated using the standard engineer's theory of shear stresses in beams and the two shear stresses added together. Account must be taken of the dissimilar materials in a strengthened beam using a modular ratio approach:

$$\tau = \frac{V}{b_a} \frac{E_f A_f y_{g1}}{(EI)_1} \tag{5.48}$$

V is the flexural shear force acting on the strengthened section and $(EI)_1$ is the flexural stiffness of the strengthened section. This component of shear stress is frequently low in comparison with the stress due to geometric discontinuity.

Step 3 – Calculate the maximum adhesive peel stress

Lack-of-fit curvature, $\Delta \psi_{fs}$

The lack-of-fit curvature is given by the difference in curvature across the adhesive joint, if no adhesive bond was present. The curvature can vary with x (along the joint).

$$\Delta \psi_{fs} = \psi_{fR} - \psi_{sR} \tag{5.49}$$

Flexibility parameters

The peel flexibility of the adhesive layer is given by:

$$a_1 = \frac{t_a}{E_a b_a} \tag{5.50}$$

The relative peel flexibility of the FRP and the metallic member is:

$$a_2 = \frac{1}{E_f I_f} + \frac{1}{E_s I_s} \tag{5.51}$$

$$a_3 = \frac{z - y_f}{E_s I_s} - \frac{y_f}{E_f I_f} \tag{5.52}$$

$$\beta = \left(\frac{a_2}{4a_1} \right)^{0.25} \tag{5.53}$$

Moment due to lack of fit and axial force in the strengthening.

$M_{PS}{}^*$ describes the moment in the beam due to the lack-of-fit between the FRP and the substrate (if there was no bond) and is a function of position, x, along the beam.

$$M_{PS}^* = C_4(x) + C_5 e^{-\lambda x} \qquad (5.54)$$

where:

$$C_4 = -\frac{1}{a_2}\left(\Delta\psi_{fs}(x) + a_3 \frac{\Delta\varepsilon_{fs}(x)}{f_2}\right) \qquad (5.55)$$

$$C_5 = \frac{a_3}{a_1\lambda^4 + a_2}\left[N_f\big|_{x=0} + \frac{\Delta\varepsilon_{fs}\big|_{x=0}}{f_2}\right] \qquad (5.56)$$

Maximum adhesive peel stress

The maximum peel stress, σ_{max}, is calculated using Equation A5.37 in Appendix 5:

$$\sigma_{max}' = -\frac{1}{b_a}\left[C_5\lambda^2 - 2C_3\beta^2\right] \qquad (5.57)$$

The boundary condition constants, C_2 and C_3, are given by the conditions at the position of the discontinuity, $x = 0$. At the end of a strengthening plate:

$$C_2 = -M_{PS}^*\big|_{x=0} \qquad (5.58)$$

$$C_3 = -\frac{1}{\beta}\left[\frac{dM_{PS}^*}{dx}\bigg|_{x=0} - \beta C_2\right] \qquad (5.59)$$

Step 4 – Calculate the principal stress, and compare with the strength of the adhesive

Calculate the maximum principal stress, σ_1, in the adhesive joint corresponding to τ_{max} and σ_{max} using Equation 5.38. Compare this with the characteristic strength of the adhesive system, $\overline{\sigma}$, which was determined in Step 1.

Step 5 – If $\sigma_1 > \overline{\sigma}$, investigate stress reduction methods

If the maximum principal stress in the joint exceeds the adhesive characteristic strength, the arrangement of the FRP plate or the joint must be modified so as to reduce the peak stresses.

Stress reduction measures

From Equation 5.47, the *maximum shear stress* can be reduced by:

- locating the plate end in a region of low beam strain
- increasing the width of the plate
- increasing the shear flexibility of the adhesive f_1, by adopting an adhesive having a lower shear modulus, or by increasing the thickness of the adhesive (with the proviso given below)
- selecting an FRP plate whose coefficient of thermal expansion is closer to that of the substrate.

A high prestress level in the plate, N_{f0}, produces high adhesion stresses at the end of the adhesive layer, unless the plate is anchored by an alternative means (such as a mechanical clamp). Similarly, a high load ($N_{f1}^{x=0}$) in the strengthening material at the end of the adhesive layer, induces high adhesion stresses, hence the tendency of the adhesive to fail adjacent to a crack in the substrate.

From Equation 5.57, the *maximum peel stress* can be reduced by:

- locating the plate end in a region of low beam moment
- increasing the width of the plate
- reducing the thickness of the plate and adhesive layer, or tapering the end of the plate
- increasing the flexibility of the adhesive a_1 (Equation 5.50)
- increasing the flexibility of the plate a_2 (Equation 5.51).

As the thickness of the adhesive layer (t_a) increases, the maximum shear stress reduces, but the maximum peel stress increases. Thus an optimum thickness for the adhesive can be calculated. A thicker adhesive joint will have a lower strength. Advice on the optimum thickness should be sought from the adhesive or strengthening system supplier, but some guidance is given in Section 7.4.2 of this report.

Details to reduce the stress concentration at the end of a plate are discussed in Section 6.6.2.

Numerical adhesive joint analysis

The adhesive joint analysis given above applies to a strengthened member of constant cross-section. When the properties of the member or strengthening plate vary along the member, it is possible in certain cases to derive a closed-form solution for the shear and peel stress in the adhesive joint. For example, Baker (1996) reports an analysis by Chalkly (1993) for tapered plates. As this work considered aerospace applications in which the composite is applied to both faces of a membrane element it may not be directly applicable to many civil engineering applications.

Where the strengthened section varies along the adhesive joint (for example, due to tapered strengthening), a numerical solution of the adhesive joint analysis is necessary. The elastic analysis equations (Appendix 5) can be solved numerically using a finite-difference procedure.

The adhesive joint is split into discrete portions along its length. For each segment of the joint, the following are calculated:

- geometric properties (for example, the thickness of the FRP)
- applied load resultants (N_s, M_s)
- lack-of-fit strain and curvature ($\Delta\varepsilon_{fs}$ and $\Delta\psi_{fs}$)
- axial force particular solution (N_{PS}) and the axial force in the strengthening (N_f)
- shear stress in the adhesive (τ)
- bending moment particular solution (M_{PS}^{*}) and the bending moment in the strengthening (M_f^{*})
- first and second differentials of M_{PS}^{*} (dM_{PS}^{*}/dx and $d^2M_{PS}^{*}/dx^2$)
- peel stress in the adhesive (σ).

The shear and peel stress distributions can then be examined to find the peak adhesive stress.

Denton (2001) also outlines a simple numerical procedure that could be used to determine stress distribution in an adhesive joint, and Frost, Lee and Thompson (2003) give an alternative solution for a tapered FRP plate. A finite element model of the adhesive joint could be used, but is unlikely to be necessary in most cases.

5.3.2 Fracture mechanics analysis of the adhesive joint

The shear and peel stresses determined from the elastic analysis method described in Section 5.3.1 are dependent upon the thickness and shear modulus of the adhesive. These are generally not uniform along the beam and are not known in a deterministic sense. Furthermore, they do not take into account the presence of bond defects, which therefore need to be allowed for in the strength parameter used.

An alternative method for assessing the ultimate capacity of the joint is a fracture mechanics approach. Fracture mechanics considers the energy released during the propagation of a crack along the adhesive joint. The fracture energy release rate (G_I) during propagation of the crack is assessed and compared with the critical energy release rate ($\overline{G}_{IC}$) required to drive the crack propagation. If $G_I > \overline{G}_{IC}$, the crack will propagate along the adhesive joint in an unstable manner, otherwise the joint is stable.

This approach is simpler to use than the elastic analysis, as it does not require the adhesive stress distributions to be determined. Fracture toughness methods inherently take account of the existence of stress discontinuities.

For the simple case of an FRP strengthening plate ending at a point of the beam where the bending moment is M, the fracture energy release rate is given by:

$$G_I = \frac{M^2}{2b_a}\left(\frac{1}{(EI)_s} - \frac{1}{(EI)_1}\right) \tag{5.60}$$

where:

$(EI)_s$ is the flexural stiffness of the unstrengthened section

$(EI)_1$ is the flexural stiffness of the composite, strengthened section

Failure occurs if:

$$G_I > \overline{G}_{IC} \tag{5.61}$$

$\overline{G}_{IC}$ is the critical energy release rate which characterises the strength, or more appropriately toughness, of the adhesive joint. $\overline{G}_{IC}$ can be determined using standard fracture mechanics tests, such as the double cantilever beam (DCB) test (BSI, 2001).

This approach does not require the thickness of the adhesive layer, and the equations are far simpler than those for the elastic analysis. Appendix 7 describes the fracture mechanics approach to assessing the ultimate strength of the adhesive layer in greater detail, including the effects of differential thermal expansion.

ANALYSIS OF AN FRP MEMBRANE SUBJECTED TO BIAXIAL LOADING

The preceding sections of this chapter have considered only strengthening for monoaxial loading. FRP can also be used to strengthen a metallic component subjected to biaxial load. For example, Figure 5.12 shows FRP applied to the web of a beam to increase the local buckling capacity of the web. The continuous fibres are orientated at ± 45° to the longitudinal axis of the member.

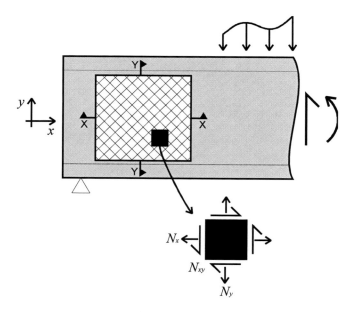

Figure 5.12 *FRP used to strengthen the web of a beam*

5.4.1 Sectional analysis

Sectional analysis is based upon the assumption that once the adhesive has cured the steel substrate and the FRP plating act together compositely. Hence the strengthened plate is characterised by composite stiffness properties of the metallic substrate and the FRP. As in the case of beam sectional analysis, it is important to take into account the strain in substrate at the time the FRP cures as an initial strain.

The FRP plating is characterised by the following constitutive relationship:

$$\begin{bmatrix} \mathbf{N} \\ \mathbf{M} \end{bmatrix} = \begin{bmatrix} \mathbf{A}_1 & \mathbf{B}_1 \\ \mathbf{B}_1 & \mathbf{D}_1 \end{bmatrix} \begin{bmatrix} \boldsymbol{\varepsilon} \\ \boldsymbol{\psi} \end{bmatrix} \tag{5.62}$$

where:

$$\mathbf{N} \equiv \begin{bmatrix} N_x \\ N_y \\ N_{xy} \end{bmatrix}, \quad \mathbf{M} \equiv \begin{bmatrix} M_x \\ M_y \\ M_{xy} \end{bmatrix} \tag{5.63}$$

are the shell membrane and flexural stress resultants per unit width

$$\boldsymbol{\varepsilon} \equiv \begin{bmatrix} \varepsilon_x \\ \varepsilon_y \\ \gamma_{xy} \end{bmatrix}, \quad \boldsymbol{\psi} \equiv \begin{bmatrix} \psi_x \\ \psi_y \\ \psi_{xy} \end{bmatrix} \tag{5.64}$$

are the shell membrane and flexural strain components.

$\mathbf{A}_1$, $\mathbf{D}_1$, $\mathbf{B}_1$ are the membrane stiffness matrix, flexural stiffness matrix, and coupling stiffness matrix for the strengthened section. These are covered in further detail in books on laminate analysis (for example, Clarke, 1996)

The stiffness matrices of the composite, strengthened plate are calculated by summing the stiffness matrices for the substrate and FRP elements, for example:

$$\mathbf{A}_1 = \mathbf{A}_s + \mathbf{A}_f \tag{5.65}$$

The membrane stiffness matrix for the metal is:

$$\mathbf{A}_s = \begin{bmatrix} E_{xs}{}^* & E_{yxs}{}^* & 0 \\ E_{xys}{}^* & E_{ys}{}^* & 0 \\ 0 & 0 & G_s \end{bmatrix} t_s = \begin{bmatrix} \dfrac{E_s}{1-v_s^2} & \dfrac{v_s E_s}{1-v_s^2} & 0 \\ \dfrac{v_s E_s}{1-v_s^2} & \dfrac{E_s}{1-v_s^2} & 0 \\ 0 & 0 & G_s \end{bmatrix} t_s \tag{5.66}$$

The membrane stiffness matrix for the FRP is:

$$\mathbf{A}_f = \begin{bmatrix} E_{xf}{}^* & E_{yxf}{}^* & 0 \\ E_{xyf}{}^* & E_{yf}{}^* & 0 \\ 0 & 0 & G_f \end{bmatrix} t_f = \begin{bmatrix} \dfrac{E_{xf}}{1-v_{xyf}v_{yxf}} & \dfrac{v_{yxf}E_{xf}}{1-v_{xyf}v_{yxf}} & 0 \\ \dfrac{v_{xyf}E_{yf}}{1-v_{xyf}v_{yxf}} & \dfrac{E_{yf}}{1-v_{xyf}v_{yxf}} & 0 \\ 0 & 0 & G_f \end{bmatrix} t_f \tag{5.67}$$

The shell strain and stress resultant components remote from the edge of the membrane can be calculated for any given loading pattern on the basis of a plate or shell model of the strengthened plate. If the structure is statically determinate, the stress resultants can be calculated by statics alone and the strains then determined by means of the above constitutive relationship for the strengthened plate. If the structure is statically indeterminate, then its response must be calculated by available closed-form solutions or numerical analysis.

The strains in the substrate and FRP are calculated from the shell strain components, taking care to subtract the strains in the substrate at the time of cure from the strains in the FRP. These strains are then compared with allowable values using the Tsai-Hill criterion (Section 3.3.3). If the strains are not within the allowable limits, then either the thickness or laminate lay-up of the FRP need to be modified. Similarly, the properties of the strengthening need to be modified if the buckling capacity, deflections, or natural frequencies fall outside allowable limits.

5.4.2 Adhesive joint analysis

Concentrated shear and peel stresses will arise in the adhesive near the edges of biaxial FRP strengthening material, just as in the monoaxial loading case.

The elastic adhesive joint analysis equations in Section 5.3.1 can be used to calculate the adhesive stresses in the edge zone of the FRP. However, the uniaxial stiffnesses (E) in

Equations 5.39 to 5.59 must be replaced by the effective stiffnesses under biaxial loading (E^*), given by Equations 5.66 and 5.67.

For example, to find the maximum adhesive shear stress along section X–X in Figure 5.12, the lack-of-fit strain (Equation 5.43) is calculated on the basis of the stress resultants parallel to the x-axis. This should be repeated for stress resultants parallel to the y-axis, to find the adhesive stress distribution along section Y–Y.

This simplified approach is less accurate in the corner regions of the plate. Finite element analysis should be used to determine the adhesive stresses in these regions. Further research work is required to extend the uniaxial adhesive joint analysis methodology presented in Appendix 5 to the biaxial case.

5.5 BIBLIOGRAPHY

Baker, A (1996). "Joining and repair of aircraft composite structures". *Mech engg trans*, vol ME21, nos 1 and 2

Chalkly, P D (1993). *Mathematical modelling of bonded fibre-composite repairs to metals*. Aeronautical Research Laboratory Research Report AR-008-365, Department of Defence and Technology Organisation, Australia

Clarke, J L, ed (1996). *Structural design of polymer composites. Eurocomp design code and handbook*. E & FN Spon, London

Concrete Society (2000). *Design guidance for strengthening concrete structures using fibre composite materials*. Technical Report 55, Concrete Society, Crowthorne

Denton, S R (2001). "Analysis of stresses developed in FRP plated beams due to thermal effects". In: J G Teng (ed), *Proc int conf FRP composites in civ engg (CICE 2001), 12–15 Dec, Hong Kong*, pp 527–536

Frost, S, Lee, R J and Thompson, V K (2003). "Structural integrity of beams strengthened with FRP plates – analysis of the adhesive layer". In: *Proc struct faults and repair 2003, London*

Highways Agency (2001a). *The assessment of highway bridges and structures*. BD 21/01 (DMRB vol 3, sec 4, pt 3), Stationery Office, London

Liu, X, Silva, P F and Nanni, A (2001). "Rehabilitation of steel bridge members with FRP composite materials". In: *Proc 1st int conf composites in construction (CCC 2001), 10–12 Oct, Porto, Portugal*, pp 613–617

Mays, G C and Hutchinson, A R (1992). *Adhesives in civil engineering*. Cambridge University Press

Moy, S S J (1999). "A theoretical investigation into the benefits of using carbon fibre reinforcement to increase the capacity of initially unloaded and preloaded beams and struts". Paper prepared under Link & Surface Transport Programme, Carbon fibre composites for structural upgrade and life extension – validation and design guidance. Dept of Civil & Environmental Engg, Univ Southampton

Moy, S S J, ed (2001b). *FRP composites – life extension and strengthening of metallic structures*. ICE design and practice guide, Thomas Telford, London

Moy, S S J, Barnes, F, Moriarty, J, Dier, A F, Kenchington, A, Iverson B (2000). "Structural upgrade and life extension of cast iron struts and beams using carbon fibre reinforced composites". In: A G Gibson (ed), *Proc 8th int conf fibre reinf composites, FRC 2000 – composites for the millennium, 13–15 Sep, Univ Newcastle-upon-Tyne*

Moy, S S J, Nikoukar F (2002). "Flexural behaviour of steel beams reinforced with carbon fibre reinforced polymer composite". In: *Proc ACIC 2002, inaug int conf use of advanced composites in construction, 15–17 Apr, Univ Southampton*. Thomas Telford, London

Sen, R, Liby, L and Mullins, G (2001). "Strengthening steel bridge sections using CFRP laminates". *Composites, Part B: Engineering*, vol 32, no 4, pp 309–322

Sen, R, Liby, L, Spillett, K and Mullins, G (1995). "Strengthening steel composite bridge members using CFRP laminates". In: L Taerwe (ed), *Proc 2nd int symp non-metallic (FRP) reinf for conc strucs (FRPRCS-2), Aug, Ghent*. E & FN Spon, London, pp 551–558

British Standard

BS 7991:2001. *Determination of the mode I adhesive fracture energy, GIC, of structural adhesives using the double cantilever beam (DCB) and tapered double cantilever beam (TDCB) specimens*

6 Design and detailing

This chapter provides guidance for the detailed design of externally bonded FRP strengthening for metallic structures. It describes the design philosophy and aspects that must be considered during design and provides the design framework within which the analysis methods described in Chapter 5 are used. It is split into the following sections:

6.1   Limit states and factors of safety

6.2   Environmental actions (hygrothermal effects)

6.3   Sizing of the strengthening material and adhesive joint

6.4   Dynamic and fatigue loading

6.5   Fire protection

6.6   Detailing

Figure 6.1 summarises the critical stages in the detailed design process, with references to the relevant report sections. Table 6.1 lists the principal parameters that are required for design. These can be used as checklists during design.

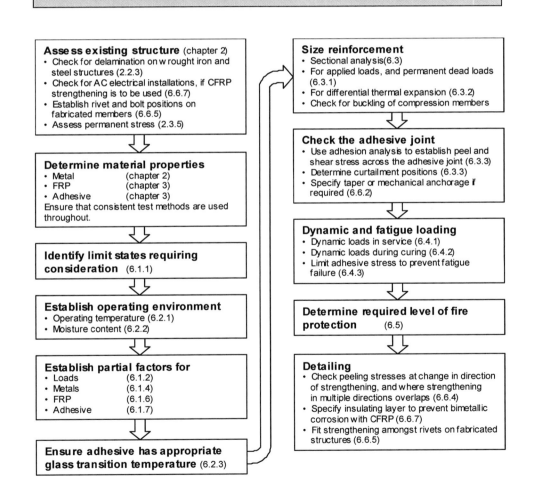

Figure 6.1   *Stages in the detailed design process*

Table 6.1 *The principal parameters required for design*

Metallic substrate	• Geometry
	• Young's modulus
	• Stress-strain curve
	• Strength or strain capacity
	• Coefficient of thermal expansion
FRP strengthening material	• Moduli of elasticity
	• Strength or strain capacity
	• Coefficient of thermal expansion
	• Glass transition temperature
Adhesive	• Shear modulus
	• Shear strength
	• Adhesive thickness
	• Glass transition temperature
	• Creep behaviour
Loading	• Load at the time of application of the strengthening
	• Temperature at the time of strengthening
	• Design loading
	• Operating temperature and environment

6.1   LIMIT STATE APPROACH AND FACTORS OF SAFETY

A coherent limit state approach should be used to design the FRP strengthening scheme. The limit state design methodology has been described in CIRIA Report 63 and is the basis of most current structural design codes, including BS 5950:2000 for steel building structures, BS 5400-1:1988 for bridges and BS EN 1990:2002 *Eurocode: basis of structural design*.

Limit state design seeks to ensure that for any given limit state the calculated effects of destabilising loads (S) are less than the resistance of the structure (R) by a margin commensurate with the required probability of failure. To achieve this, design variables subject to statistical uncertainty are factored by partial safety factors. This can be expressed as:

$$S\!\left(x_{L,i}\,\gamma_{fL,i}\right)\gamma_{f3} < R\!\left(a_d,\, x_{M,i}\,/\gamma_{M,i}\right) \tag{6.1}$$

$x_{L,i}$ are the load variables, a_d are the design values of the geometric data and $x_{M,i}$ are the material variables.

The partial safety factors are:

- $\gamma_{fL,i}$ on the ith applied loads
- γ_{f3} allows for uncertainty in the design
- $\gamma_{M,i}$ on the ith material property.

For the simple case of only the effect of one load and one significant material strength parameter, this reduces to:

$$S \, \gamma_{fL} \, \gamma_{f3} < \frac{R}{\gamma_m}$$

(6.2)

6.1.1 Limit states to be considered

The limit states that require consideration can be split into a number of categories.

Ultimate limit states describe partial or complete collapse of a strengthened structure, leading to severe consequences in terms of financial or human loss. They include:

- failure of the FRP by fibre rupture
- failure of the adhesive joint, resulting in debonding of the FRP from the substrate
- failure of the substrate, by fracture, yield or local buckling
- local or global buckling failure of the strengthened member
- local buckling or delamination failure of the FRP.

Serviceability limit states describe the proper in-service performance or appearance of the structure. They include:

- limiting deflection
- avoidance of permanent set or damage due to cracking or plastic strain
- limiting vibration
- strain limits to avoid initial damage to the FRP matrix, which would allow moisture or chemical ingress, and reduce fatigue life.

Durability limit states relate to ageing or time-related effects:

- fatigue failure
- creep rupture
- corrosion
- weathering
- erosion and wear.

Hazard limit states consider risk-related unintentionally applied loads, such as:

- abnormal impact
- fire.

The relevance of these different limit states to a strengthening scheme will depend upon the structure's characteristics and the aims of the strengthening scheme. However, all limit states should be considered during design. For example, if strengthening is used to reduce the serviceability deflection of a structure, it is important to also ensure that the strengthening material does not fail before the ultimate limit state is reached.

6.1.2 Design life

The design life of the strengthened structure should recognise the intention of the strengthening scheme. For example, FRP strengthening may be required to extend the life of a fatigue-damaged bridge until the structure can be permanently removed and replaced. In this case, the design life is limited by the remaining fatigue life of components of the structure that are not being strengthened, which might be 20 years.

The required design life of the strengthened structure should be established before determining the loads acting on the structure and the partial factors on material properties (Section 6.1.4).

6.1.3   Partial factors on the loads

The loads and partial load factors γ_{fL} specified in limit state codes of practice are generally also applicable to composite and FRP-strengthened structures. For example, BD 37/01 (Highways Agency, 2001b) specifies load factors for highway bridges and BS 5950:2000 considers steel building structures.

BS 5400-1:1988) specifies values of γ_{f3}, which are also applicable to FRP-strengthened structures, whereas building codes (such as BS 5950), combine γ_{f3} and γ_{fL} into a single partial factor.

6.1.4   Partial factors on the materials

The material partial factor, γ_m, can be separated into three effects, each of which is expressed by a separate partial factor:

$$\gamma_m = \gamma_{mv}\,\gamma_{me}\,\gamma_{mt} \tag{6.3}$$

Factor to take account of material variability (γ_{mv})

γ_{mv} is related to the statistical variability of the material property and hence to the probability that the material property will exceed a reference value. γ_{mv} is equivalent to γ_m in current codes for materials that are not affected by environmental and age-related effects

Factor to take account of environmental effects (γ_{me})

The mechanical properties of composite materials are influenced by temperature and moisture conditions, referred to collectively as hygrothermal conditions. The factor γ_{me} takes these hygrothermal environmental conditions into account.

Factor to take account of age-related effects (γ_{mt})

Creep and age-dependent damage are expressed by the time-dependent factor, γ_{mt}.

Each of the components of the strengthened structure (the metal substrate, the FRP strengthening material and the adhesive bond) requires separate treatment. Section 6.1.4 covers the material partial factors for the metal substrate, the material partial factors for the FRP composite and the material partial factors for the adhesive joint.

Material partial factors for the metal substrate

The design strength of the existing metallic structure is assessed using the material partial factors given in the appropriate design codes:

*   BS 5400:1988, BS 5950:2000 and BD 56/96 (Highways Agency, 1996) for modern steel
*   BD 21/01 (Highways Agency, 2001a) for cast iron, wrought iron and historic steel.

Cast iron is conventionally designed. Conventionally, cast iron is designed using a permissible stress approach, so that explicit material partial factors are not required. Where a statistically significant number of test results are available for the strength of cast iron, Bussell (1997) suggests that a material partial factor γ_m of 3 can be applied to the characteristic strength.

Material partial factors for the FRP material

FRP strengthening materials are available in a wide range of forms, produced using a variety of manufacturing methods, from different constituent materials. Variability of the FRP depends on many factors, including voids, fibre misalignment, lack of fibre matrix interfacial bond and microcracks in the matrix.

Material variability factor

The variability of an individual material property can be determined from tests on a statistically significant number of specimens (normally 10 specimens). These tests are analysed to determine the coefficient of variation of the property, and the partial factor γ_{mv} derived using the equations given in Appendix 8.

The test samples used to determine the material variability should be cut from material produced using the same technique that will be used in the final application. In the case of preformed plates, obtaining representative samples is straightforward. Where an *in-situ* technique is used to produce the strengthening material, or the composite material is created specifically for a particular application, it will be necessary to produce samples that are as similar as possible to the final application.

In practice, a minimum of five test samples will be required to quantify the material variability, although a true statistically significant set of results requires a larger number of samples to be tested. Where representative samples, or a statistically significant number of test results are not available for analysis, a larger γ_{mv} should be adopted.

If the material variability cannot be assessed from test samples, representative values based on previous experience can be used in design. TR55 (Concrete Society, 2000) suggests partial factors for the variability of FRP materials used for strengthening concrete, which are generally the same products used to strengthen metallic structures. These factors are based upon guidance in the Eurocomp design code (Clarke, 1996).

According to TR55, the variability of the FRP (γ_{mv}) can be split into the variability of the constituent matrix and fibre materials (γ_{mf}) and the manufacturing process by which the FRP is produced (γ_{mm}):

$$\gamma_{mv} = \gamma_{mf} \, \gamma_{mm} \tag{6.4}$$

Values of γ_{mf}, and γ_{mm} are given in Tables 6.2 and 6.3. Although TR55 gives a partial factor of 3.5 for GFRP, factors determined by calibration of actual glass-fibre samples give a considerably lower γ_{mf}, usually not exceeding 1.5

Table 6.2 *Material partial factors for the variability of the FRP material (Concrete Society, 2000)*

Material	γ_{mf}
Carbon FRP	1.4
Aramid FRP	1.5
Glass FRP	3.5

Table 6.3 *Material partial factors for variability in the manufacturing process (Concrete Society, 2000)*

Manufacturing process and form of composite	γ_{mm}
Plates	
Pultruded	1.1
Prepreg	1.1
Preformed	1.2
Sheets or tapes	
Machine-controlled application	1.1
Vacuum infusion	1.2
Wet lay-up	1.4
Prefabricated (factory-made) shell	
Filament winding	1.1
Resin transfer moulding	1.2
Hand lay-up	1.4
Hand-held spray application	2.2

Environmental factor

At elevated temperatures, the temperature dependence of a material property, X, is expressed by a simple power law function. This can be used to determine the value of the environmental safety factor, γ_{me} for elevated temperatures.

$$X(T) = X(T_{ref}) \left[\frac{T_g - T}{T_g - T_{ref}} \right]^n \tag{6.5}$$

n is an experimental exponent, normally taken to be 0.5

T is the operating temperature

T_{ref} is the reference temperature, for which the material property is known

T_g is the glass transition temperature of the matrix. The glass transition temperature is a function of the moisture content of the matrix (see Section 6.2).

The simplifying assumption is made that the mean value of the matrix property is a function of the hygrothermal conditions, but the coefficient of variation is not affected.

The elastic modulus of the FRP matrix at a given temperature, $E_{matrix}(T)$, is derived from the reference modulus $E_{matrix}(T_{ref})$ using:

$$E_{matrix}(T) = E_{matrix}(T_{ref}) \sqrt{\frac{T_g - T}{T_g - T_{ref}}}$$ (6.6)

The longitudinal modulus of FRP material with unidirectional fibres is determined using the rule of mixtures (Equation 3.1):

$$E_f(T) = E_{fibre} V_{fibre} + E_{matrix}(T)(1 - V_{fibre})$$

The environmental factor for stiffness is then determined from the change in modulus due to a temperature change from the reference temperature, T_{ref}, to the operating temperature T:

$$\gamma_{me} = \frac{E_f(T_{ref})}{E_f(T)}$$ (6.7)

A similar approach is used to assess the environmental safety factor for material strength.

The material properties of an FRP are also reduced low temperatures. For example, Figure 3.3 shows that embrittlement of the resin occurs at low temperatures. Tests should be carried out to characterise the low-temperature performance of the FRP and adhesive and hence determine a environmental safety factor appropriate to low-temperature performance.

Time-related factor

Two mechanisms result in the change in mechanical properties of a material with time:

- creep
- material degradation

Creep results from the long-term sustained stress within the material. In an FRP material, the matrix material is susceptible to creep. For unidirectional laminates, however, the longitudinal properties of the composite are dominated by the fibres, so creep in the fibre direction is low.

The time-related partial factor should ideally be determined from accelerated ageing tests on the FRP material being used (Griffith *et al*, 1980). The material supplier should provide data on the long-term strength and modulus retention of their product, which is used to determine the time-related factor.

Alternatively, Findley's law (described in Section 3.4.3) can be used to estimate creep strain as a function of time, but this does not take into account degradation of the material (Findley, 1971). The total elastic and creep strain at time t elapsed from the application of a permanent stress σ_t is:

$$\varepsilon(t, \sigma_t) = \varepsilon_0 + j t^k$$ (6.8)

where:

$\varepsilon(t, \sigma_t)$ is the strain at time t, due to the application of stress σ_t

ε_0 is the time-independent strain (dependent upon the applied permanent stress, σ_t)

j is a coefficient that depends upon the applied permanent stress

k is a constant that is independent of time and stress but that may be a function of environmental conditions (temperature, moisture content etc).

The coefficients j and k must be determined experimentally.

The effective modulus allowing for creep is defined by:

$$E_{eff}(t, \sigma_t) = \frac{\sigma_t}{\varepsilon(t, \sigma_t)} \qquad (6.9)$$

so that

$$E_{eff}(t) = \frac{E_0}{1 + \dfrac{E_0}{\sigma_t} jt^k} \qquad (6.10)$$

where:

$$E_0 = \frac{\sigma_t}{\varepsilon_0} \qquad (6.11)$$

The reduction in matrix modulus with time can be found, and the reduction in composite stiffness found using the rule of mixtures (Equation 3.1).

Eurocomp (Clarke, 1996) includes a combined partial factor for both environmental and time effects. The values in Table 2.6 of Eurocomp can be used for guidance, but they do not allow environmental and time-related effects to be separated.

Material partial factors for the adhesive

Material variability factor

The variability of the adhesive joint strength is dependent upon:

- variations in thickness, shear modulus, and shear strength
- the presence of voids, disbonds and other random defects
- variability in the interlaminar shear strength of the FRP and the substrate.

There appears to be no authoritative quantitative guidance on the variability of the adhesive strength when applied to a strengthening application. As with the FRP, a series of tests would need to be undertaken to obtain an accurate assessment of the variability of the strength of an adhesive joint.

For guidance purposes and preliminary design, the partial factors proposed in Eurocomp (Clarke, 1996) for an adhesive joint can be used.

The material variability factor is subdivided into two components:

- γ_{mf}, which depends upon the certainty of the value used
- γ_{mm}, which depends upon the method of application.

$$\gamma_{mv} = \gamma_{mf}\,\gamma_{mm} \qquad\qquad (6.4)$$

Values for γ_{mf} and γ_{mm} are given in Tables 6.4 and 6.5.

Table 6.4 *Material partial factors for variability of the adhesive joint due to the method of characterisation (Clarke, 1996)*

Source of the adhesive properties	γ_{mf}
Typical or textbook values (for appropriate adherends)	1.5
Values obtained by tests on representative specimens	1.25

Table 6.5 *Material partial factors for variability of the adhesive joint due to the method of application (Clarke, 1996)*

Method of adhesive application	γ_{mm}
Manual application, no adhesive thickness control	1.5
Manual application, adhesive thickness controlled	1.25
Established application procedure with repeatable and controlled process parameters	1.0

Environmental factor

Adhesives are more sensitive to hygrothermal conditions than FRP strengthening materials. Their glass-transition temperature is lower, and there are no fibres to mitigate the hygrothermal effects on the resin.

The effects of temperature and moisture upon the adhesive material properties can be assessed using Equation 6.5. As there are no fibres to be considered, the partial environmental factor can be found directly, using:

$$\gamma_{me} = \sqrt{\frac{T_g - T_{ref}}{T_g - T}} \qquad\qquad (6.12)$$

where:

    T is the operating temperature

    T_{ref} is the reference temperature, for which the material property is known

    T_g is the glass transition of the adhesive (see Section 6.2).

For comparison purposes, Eurocomp (Clarke, 1996) suggests the partial environmental factors in Table 6.6.

Table 6.6 *Material partial factors for environmental conditions (Clarke, 1996)*

Environmental conditions	γ_{me}
Service conditions outside test conditions	2.0
Adhesive properties determined for the service conditions	1.0

Time-related factor

An adhesive joint is more susceptible to creep than the FRP due to the absence of creep-resistant fibres. However, the creep in the adhesive will be negligible unless permanent stresses are applied as a result of prestressing, load-relief jacking or superimposed dead loads. Creep can be beneficial within the adhesive joint, since it reduces the stiffness of the adhesive. This results in an increased transmission length near the end of a strengthening plate and consequently a reduction in the peak stresses.

Creep will be most significant in regions of high bond stress, such as near the ends of a prestressed strengthening plate (see Section 5.3.1).

As in the case of FRP, the material supplier should provide information on the time-dependent properties of the adhesive, including both creep and degradation of the adhesive, which can be used to assess γ_{mt}. Property degradation should be assessed from accelerated ageing tests.

Time-related partial factors for an adhesive joint are given in Table 6.7. The short-term loading factor is taken from Eurocomp. For long-term loading, Eurocomp gives $\gamma_{mt} = 1.5$, however; a higher value of $\gamma_{mt} = 2.0$ is suggested here, as design is dominated by the peak stresses within the adhesive joint (Section 5.3.1) and experimental validation of the design approach is required (Appendix 5).

Table 6.7 *Material partial factors for length of loading*

Type of loading	γ_{mt}
Long-tern loading	2.0
Short-term loading	1.0

6.2 ENVIRONMENTAL ACTIONS

The operating environment is particularly important for an FRP-strengthened structure. It must be considered at two stages during design.

1. The operating temperature and humidity affect the material properties of the adhesive and FRP, so they are needed to determine the environmental partial factors (using Equation 6.5).

2. The effects of differential thermal expansion between the metallic substrate and the FRP strengthening material must be assessed, as these can dominate the design of the adhesive joint (see Section 6.3.1)

6.2.1 Assessing the operating temperature

The temperature distribution in a strengthened structure operating within an environment having a given temperature may, in theory, be determined from first principles by a heat flow analysis, given the environmental temperature, the coefficients of thermal conductivity of the materials and the surface heat transfer coefficients. This approach is onerous and subject to uncertainty.

A more practical approach is to use code guidelines. The environmental temperature for bridges in the UK is specified in BD 37/01 (Highways Agency, 2001b). The standard also provides temperature distributions in common forms of highway bridge deck that obviate the need for a heat flow analysis in the structure, except for a local adjustment to take into account the presence of the strengthening material. In outline, the BD 37/01 methodology involves the following steps.

1. Determine the extreme maximum and minimum shade air temperatures at the location of the structure (using maps of isotherms for the UK).

2. Find the maximum and minimum bridge temperatures from the air shade temperatures, depending upon the structural form of the bridge and the thickness of applied surfacing.

3. Deduce the temperature distribution through the depth of the bridge.

Given that the strengthening material is to be applied to an existing structure, it is advisable to take measurements of existing temperatures on the faces of the structure that will receive the strengthening.

If the strengthening scheme is unlikely to cause much disturbance to heat transfer between the bridge and the surrounding air (such as a local strengthening material, applied to part of the soffit of a beam), the temperature distribution in the unstrengthened structure should be used to determine the operating temperature within the composite strengthening.

When the FRP covers a larger area of the metal structure, it may affect the temperature distribution in the structure. In this case, the heat flow and temperature distribution should be determined, taking into account the interaction of structure and strengthening. Table 6.8 indicates typical thermal conductivities for the relevant materials.

Table 6.8 *Thermal conductivities of the substrate and strengthening materials*

Material	Thermal conductivity (W/m.K)
Cast iron	55
Steel	50
Carbon fibre	17
Aramid fibre	4
Glass fibre	1
Epoxy adhesive	0.8

It should be noted that the thermal conductivity properties of certain types of fibre and of FRP are anisotropic. If necessary, the composite properties can be predicted from the conductivities of the constituent materials using formulas that can be found in the literature. The conductivity of the matrix is significantly affected by any filler. It is recommended that if the thermal conductivity is an important design parameter, then it should be established by tests.

The temperature distribution within the structure also depends upon the heat transfer coefficient of the surfaces, and hence upon any coatings applied. For example, painting the surface white will alleviate adverse temperature problems. There appears to be no reliable data available for the effect of surface coatings, however.

Glass transition temperature of the resins and adhesives

The glass transition temperature is defined as the temperature at which the adhesive or FRP matrix changes from a solid to a viscous state, with consequential reduction in stiffness and strength (Figure 3.2). The glass transition temperature T_g is the approximate midpoint temperature for the change in material state. As discussed in Section 3.1.2, the glass transition temperature is a property of the resin that also depends upon the temperature during curing of the adhesive.

The glass transition temperature should be obtained from the material supplier. It will usually be the glass transition temperature of the adhesive, rather than the matrix, that governs design, especially when preformed plates (cured at elevated temperature) are bonded to the structure using an ambient cure adhesive.

The value of the glass transition temperature is to some extent dependent upon the test method used to obtain it (see Section 7.1.1). It is vital to ensure that all glass transition temperature data used in a design is obtained using the same test method.

Moisture dependence

Absorption of moisture produces an apparent reduction in the glass transition temperature, hence the moisture dependence of the matrix and adhesive material properties can be modelled as a shift in T_g:

$$T_g(m) = T_g - m\,\Delta T_g \qquad (6.13)$$

The moisture content m is expressed as a fraction of the maximum moisture content and ΔT_g is the shift for 100 per cent saturation. ΔT_g must be determined for a particular material, and should be provided by the materials supplier. For guidance, ΔT_g for an e-glass/polyester pultruded plate is around 13°C. The maximum moisture content of a composite can be determined using the test method given in BS EN 2378:1994.

The modified glass transition temperature is used in conjunction with Equation 6.5 to assess the reduced material properties of the matrix or adhesive, and thereby determine the environmental partial factor.

Operating temperature limit

There is significant loss of resin strength at temperatures approaching the glass transition temperature, as is shown in Figure 3.2. It is common practice to ensure that the maximum operating temperature (T) is below the glass transition temperature (T_g) by a margin of at least T_A. This criterion will usually dictate the choice of adhesive.

$$T_g - T \geq T_A \qquad (6.14)$$

Depending upon the application, T_A should be in the range of 10–20°C.

A more rational method is to use the limit state approach. The operating temperature is treated as an action, and is thus factored by a partial load factor γ_{fL}, while the glass transition temperature, being a material property, is factored by a material partial factor γ_m, describing its statistical variability (Section 6.1.4). The design requirement becomes:

$$T\,\gamma_{fL} \leq \frac{T_g}{\gamma_m} \qquad (6.15)$$

6.3   SIZING OR CHECKING THE STRENGTHENING ELEMENTS

There are two stages to sizing or checking the elements of a strengthening system.

1. Size or check the FRP strengthening material at the section of maximum normal stress in the FRP, using sectional analysis.

2. Check the strength of the adhesive joint at the ends of the FRP or at other discontinuities such as steps in thickness or cracks in the substrate, using an adhesive joint analysis.

6.3.1   Differential temperature effects

Differential thermal expansion can lead to significant stress concentrations in the adhesive near the ends of the strengthening material. Observations during and after the installation of UHM CFRP on a cast iron bridge have indicated that the strain variation due to daily temperature cycles is greater than that due to the applied vertical loads.

Differential thermal expansion must be considered in both the sectional analysis and the adhesive joint analysis of the strengthened structure. Methods for calculating the effects of temperature on an FRP-strengthened metallic structure are described in the following sections of the report:

*   the range of operating temperatures, in Section 6.2

*   the effects of temperature on the sectional analysis, in Section 5.2

*   the effects of temperature on the adhesive joint analysis, in Section 5.3.

If necessary, adhesive stress concentrations due to temperature effects can be reduced by such measures as tapering the end of the strengthening material (Section 6.6.2). It is also worth noting that the peak differential temperature stresses occur at the end of plates, where the stresses due to permanent and live loads are generally low.

Stress-free temperature

The temperature at which the adhesive cures defines the stress-free temperature for the adhesive layer and the strengthening material, which will usually be higher than the structure's operating temperature. Restrained differential thermal expansion due to cooling of the strengthening system results in locked-in stresses in the FRP, substrate and adhesive layer.

The curing temperature of the adhesive is usually higher than the ambient temperature at which installation occurs due to exothermic heat generated in the adhesive during the curing reaction and the possible use of heaters to raise the temperature of the structure and strengthening materials (Section 7.2.3). The metallic structure, however, tends to act as a heat sink, and the temperature at which the adhesive sets is difficult to determine with certainty.

In a purely unidirectional composite, the transverse strength of the FRP is low and the restrained thermal stress can lead to longitudinal cracking of the strengthening material due to transverse stresses. This does not affect the short-term load capacity of the FRP, but the durability of the FRP is compromised, as the cracks expose the fibres to ingress of moisture and pollutants. Longitudinal cracking can be alleviated by the use of a properly formulated epoxy and the inclusion of an adequate quantity of transverse fibres in the composite.

6.3.2 Sectional analysis

Sectional analysis of a strengthened member is presented in Section 5.2. Factored loads and material properties should be used in this analysis to determine the quantity of FRP strengthening material required.

6.3.3 Adhesive joint analysis

The design must ensure that the peak adhesive stresses at the end of the strengthening material, or adjacent to a crack in the substrate do not exceed the strength of the adhesive joint. The peak adhesive stress is found using the adhesive analysis presented in Section 5.3, in particular Equations 5.47 and 5.57. The strength of the adhesive joint is governed by the weakest component of the joint, which may be:

- the adhesive
- the FRP (delamination failure)
- the substrate (for example, delamination of wrought iron)
- the adhesive-FRP or substrate-FRP interface (where adhesion occurs)

In the analysis both the load and the strength parameters should be suitably factored by appropriate partial safety factors (Section 6.1).

Various methods of reducing the stress concentration near the end of the FRP are discussed in Section 6.6.2. Where the bond stresses cannot be carried by the adhesive alone, a mechanical anchorage will be necessary, which will need to be designed for the specific application. Mechanical anchorage will usually be necessary if the FRP is prestressed and bonded to the metallic member.

For non-uniform plates or members, a numerical method of analysis is likely to be necessary. The elastic adhesive joint analysis equations in Section 5.3.1 can be solved numerically by a discretisation scheme, allowing variations in the properties of the strengthening material to be analysed. Numerical methods include the finite difference method, the finite element method, and the boundary element method. The finite element method has the advantage that elasto-plastic material behaviour can be readily taken into account, allowing stress concentrations to be reduced due to plastic redistribution. Numerical methods are in general approximate methods whose accuracy depends on adopting a sufficiently dense mesh in zones of high stress gradient such as in stress concentration regions. A numerical solution procedure to determine the adhesive stress distribution in a non-uniform strengthened member is described by Denton (2001).

6.4 DYNAMIC LOADING AND FATIGUE

6.4.1 Assessment of strength with respect to dynamic loads

FRP can be used to increase the strength of a metallic structure with respect to dynamic loading, such as impact loading.

The effects of dynamic loading should be taken into account during the stress analysis, where relevant. For example, BD 37/01 (Highways Agency, 2001b) provides guidance on dynamic loading effects for railway and highway structures and impact loads due to vehicle collision.

The material properties used for dynamic loading should be appropriate for the strain rate expected during the dynamic loading. Laboratory static tests are normally carried out at low strain rates. These are appropriate for static loading in the real structure, and are normally adequate for dynamic traffic loading, since this has a low frequency. Impact loading, on the other hand, may occur at a high strain rate, and the impact strength of the strengthened structure should be assessed on the basis of high strain rate material properties. Rather than undertake high strain rate material testing, it may be more cost-effective to determine the impact resistance of the structure by a suitable impact test.

6.4.2 Loads during curing

It is usually desirable to minimise the length of time for which a structure is closed for strengthening works. This is particularly true for railway and road bridges, where the costs associated with bridge closure can be high. The bridge might be reopened to traffic before the adhesive has fully cured, or even trafficked throughout the curing process.

Traffic loading imposes dynamic loads on the strengthened structure, and hence cyclic stresses in the partially cured adhesive layer. These cyclic stresses can have a deleterious effect on the final strength of the adhesive layer since they can be sufficiently large in relation to the strength of the partially cured adhesive to produce permanent damage in the form of internal cracks. The strength reduction is greatest near the ends of the strengthening material, where the maximum adhesive shear stress under live load occurs (see Section 5.3.1).

Measures can be taken to reduce the magnitude of the dynamic loads (for example, by placing speed and weight limits over the structure), but without closing the bridge it is not possible to eliminate the dynamic loads altogether.

Barnes and Mays (2001) conducted dynamic tests on strengthened steel beams and showed that dynamic lap-shear tests can be used to predict the reduction in adhesive strength due to dynamic loading during curing. Lap-shear samples were subjected to strain-controlled cyclical loading of constant amplitude during the curing period, following which they were tested to failure. These tests were conducted at several frequencies and yielded the reductions in strength shown in Table 6.9. As the table shows, the higher the amplitude of the imposed dynamic strain during curing, the greater the reduction in strength of the bonded joint when compared with the strength obtained when no loading is applied during curing.

Table 6.9   *Strength reduction resulting from direct strain in the metal substrate (Barnes and Mays, 2001)*

Strain amplitude during curing (due to dynamic loading) (%)	Percentage strength reduction
0.002	10
0.005	12
0.010	16
0.015	22
0.020	32

Barnes and Mays (2001) suggest that if a structure is subjected to traffic while the adhesive cures, the strain variation in the bottom-fibre of the metallic beam due to the traffic should be assessed and the adhesive strength degraded accordingly. Further research is required on this subject. Tests to investigate the effects on joint performance of dynamic loading whilst the adhesive cures are being carried out at Southampton University at the time of writing of this report (Moy, 2001b).

6.4.3   Fatigue strength assessment

The fatigue endurance should be assessed in each of the three components of an FRP-strengthened structure:

- the metallic substrate
- the adhesive bond
- the FRP.

Fatigue damage in the metallic structure

The fatigue life of the existing metallic structure can be assessed using standard design code methods (for example, BS 5400-10:1980 and BS 7608:1993).

Externally bonded FRP can be used to extend the fatigue life of existing structures by reducing the stresses to which a fatigue-sensitive detail is subjected. Prestressing the FRP so as to induce a permanent compressive stress inhibits the growth of fatigue cracks (Bassetti, 2001; Bassetti *et al*, 2000a, 2000b). In welded structures this approach is less effective due to the high residual tensile stresses induced by the welding process. A particular advantage of prestressed FRP for structures that already contain significant fatigue cracks is that the fatigue life of the whole member can be extended, rather than just that of a local detail (as would be the case with more traditional techniques such as re-welding or hole drilling).

Fatigue damage in the FRP strengthening material

Most FRP laminates have good fatigue resistance for in-plane loading parallel to the principal fibre direction (as discussed in Section 3.3.8). The fatigue resistance of the FRP laminates used in structural strengthening schemes is unlikely to be critical. The adhesive joint, however, is more susceptible to fatigue damage, and limits are placed upon the strain on the FRP strengthening material (see Table 6.10, below) to limit the fatigue action in the adhesive.

The fatigue life of a composite component can be characterised by an *S–N* curve, of the form:

$$\frac{\sigma_n}{\overline{\sigma}_n} = 1 - K \log_{10}(n_\mathrm{f}) \qquad (6.16)$$

where σ_n is the *maximum* stress in a fatigue cycle and n_f is the number of fatigue cycles to failure. $\overline{\sigma}_n$ is the static failure stress corresponding to σ_n. The slope parameter K depends on the type of matrix and fibre and on the R ratio, which is the ratio of minimum to maximum stress in a cycle. K is usually 0.1 or less for FRP. Typical values of K are shown in the *S–N curves* in Figure 6.2.

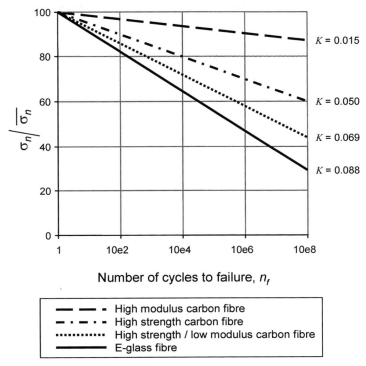

Figure 6.2 S–N *diagrams for representative unidirectional FRP materials showing the effect of fibre type and modulus for tension-tension fatigue (adapted from Jones, 1994)*

Fatigue damage in the adhesive joint

The adhesive joint is more susceptible to fatigue failure than the FRP strengthening material. A properly made adhesive joint, however, can have superior fatigue characteristics to an equivalent riveted joint.

The zones of stress concentration at either end of a strengthening plate will be most susceptible to fatigue, but the live load stress cycles in the adhesive are generally small if the plate ends are located in a zone of low stress in the beam. High stress concentrations also arise adjacent to cracks in the substrate. If there is a possibility of a crack in the substrate in the zone of maximum flexural stress, then the fatigue issue should be carefully investigated.

Stresses due to temperature variations can be high, but the number of cycles of such stresses is considerably lower than the number of live load stresses, and the rate of loading is slow.

S–N curves to characterise the adhesive joint in an externally bonded FRP strengthening system are not currently available. However, tests carried out by the National Physical Laboratory on bonded joints suggest that the *S–N* curve formulation for FRP, Equation 6.16 and Figure 6.2, is also applicable to adhesive joints. This *S–N* curve suggests that the peak shear stress in a fatigue cycle should not exceed 20–30 per cent of the ultimate static failure strength.

In the absence of experimental *S–N* curves for the peak stress in the adhesive joint, stress or strain limits have been proposed to avoid fatigue damage. If the strain in the adhesive layer under full live loading at the serviceability limit state is kept below that at first cracking (corresponding to the first "knee" in the load-deflection curve), then fatigue failure without forewarning is relatively unlikely. TR55 (Concrete Society, 2000) suggests that the stress range in the FRP, when applied to a concrete structure, should be limited to a proportion of the composite's design ultimate strength, in accordance with Table 6.10. This follows the approach suggested in BA 30/94 (Department of Transport, 1994) for steel-bonded strengthening of concrete structures, and aims to limit the stress in the adhesive.

Table 6.10 *Maximum cyclic stress range suggested in TR55 (Concrete Society, 2000)*

Composite material	Permissible stress range as a proportion of the design ultimate strength (%)
CFRP	80
AFRP	70
GFRP	30

The high permissible stress ranges quoted for CFRP and AFRP are unlikely to be ever attained in practice, because of the material safety factors specified for static loading. Fatigue assessed on this basis is unlikely to control the design, therefore. In most cases, the fatigue stress range to which the adhesive joint is subjected is likely to be relatively low. For example, since the peak strain that can be tolerated by a cast iron structure is very low compared with the strain capacity of the FRP, the cyclic strain in the adhesive layer will usually be low. However, since the fatigue regime is dependent on the application, it is recommended that in fatigue critical applications fatigue tests should be undertaken to characterise the adhesive joint.

6.5 FIRE PERFORMANCE AND PROTECTION

The heat produced by a fire adjacent to the strengthened structure causes the polymer components to soften, as described in Section 3.1.2. Direct exposure to flames will cause surface charring. Furthermore, the organic polymers in the composite matrix and the adhesive are themselves flammable and produce varying amounts of toxic by-products as they burn (Section 3.3.9).

It will usually be the adhesive that governs the fire performance of an FRP strengthened metallic structure, due to its lower T_g value (Section 3.4.4). The FRP will provide fire protection to the adhesive for a certain period of time depending on the FRP's thickness, but the metallic member can conduct heat directly to the adhesive joint.

A temperature gradient developed through the thickness of the composite results in a gradient in the mechanical properties of the composite and in differential thermal expansion effects. The resulting stress distribution results in bending in the strengthening material, increasing the peel stresses.

If the FRP fully covers the surface of a metallic structure exposed to a fire, it may improve the fire performance of the structure since the FRP slows down the rate of heating of the metal.

6.5.1 Fire design

When designing a strengthened structure for accidental fire load, reference should be made to the appropriate standards for the structure being strengthened (BS 5400-1:1988 and BS 5950:2000). These codes specify reduced factors of safety for use during accidental fire damage. The fire performance of cast iron and wrought iron are discussed in Sections 2.2.1 and 2.2.3.

An initial assessment of the fire protection requirements can be made on the basis of the fire protection requirements of the existing structure. An enclosed metallic building structure usually requires passive fire protection, so the strengthening system is similarly likely to require fire protection. However, a protection system designed to prevent fire damage to steel will not necessarily be the most appropriate protection for FRP. In an enclosed environment, FRP may need to receive fire protection to delay the time to the onset of smoke and toxic fume emissions. Suitable passive fire protection systems are examined in Section 6.5.3.

Conduction of heat through the metallic member being strengthened must be evaluated, as this can lead to a large temperature increase at the bonded joint. Additional fire protection will be required around adjoining metallic elements of the strengthened structure.

A structure in an open environment, such as a railway bridge, will not normally have fire protection, although it will still be susceptible to fire. The existing, unstrengthened structure may be able to support its own weight under the action of a fire for an adequate length of time. If this is the case, loss of the FRP will not lead to failure of the structure, and the local damage to the strengthening material can readily be repaired after the fire (Section 8.3). The designer should, however, be satisfied that during the fire the structure will not be required to support heavy live loads, such as fire-fighting equipment. Neither should the FRP be required to carry large permanent loads, for example where an opening is made in a floor slab after a beam has been strengthened.

The FRP strengthening material may be required to carry significant permanent loads, for example, in cases where the FRP is applied to permit the removal of props from under a bridge, or to allow a building structure to be modified. In these cases, loss of the FRP may lead to failure, and fire protection measures should be taken (Section 6.5.3).

When the adhesive has a relatively low glass transition temperature T_g, and the FRP has an adequate fire performance, mechanical anchorages can be provided at the ends of the strengthening component.

6.5.2 Fire tests

Standard test methods are available to classify the fire performance of materials (BS 476-11 to 19:1987), but these small-scale tests cannot necessarily be extrapolated to the fire performance of composites on a structural scale. The general lack of information on the effectiveness of externally bonded FRP during a fire can only be addressed by fire tests.

The general principles for determining the fire resistance and endurance of construction elements are given in BS 476-20:1987. The fire resistance is defined as the time to failure with respect to the following criteria:

- *load bearing capacity* – the ability of the specimen to support the test load without collapse and without exceeding specified magnitudes or rates of deflection
- *integrity* – the ability of a separating element to contain the fire, ie not be breached by the fire
- *insulation* – the cooler face of a separating element to remain below a safe limit.

6.5.3 Fire protection

A layer of passive fire protection can be applied over the FRP strengthening material to isolate it from the fire, and to insulate the adhesive from elevated temperature. The fire protection must also be sufficiently robust: in particular, it must not be readily eroded by fire-fighting equipment.

Fire protection options available include:

Intumescent coatings

Intumescent coatings can be applied to both steel and FRP. Gases are evolved in the coating at a specific *activation* temperature and produce a protective, charred foam around the structure.

Water-based, thin (0.4–2 mm) intumescent coatings typically give a fire resistance of between 30 and 90 minutes. These are likely to be most suitable for heritage structures, but suffer from poor impact resistance. Epoxy-based, thick (2–20 mm) coatings can provide a fire resistance of 120 minutes and are more suitable for outdoor use.

The activation temperature of the intumescent coating must be lower than the glass transition temperature of the FRP strengthening and adhesive. Intumescent paints for steel structures can have activation temperatures of 100–150°C and are unlikely to be appropriate. However, there will be a temperature gradient within the body of the FRP dependent upon its thickness and the properties of the substrate, which may result in a significantly lower temperature at a certain depth below the surface. If the time to loss of the structural capacity of the strengthening in a fire is a critical issue, it is recommended that fire tests be carried out to determine the performance of a particular combination of intumescent coating, FRP, adhesive and substrate.

Cladding

Calcium-silicate boards can be used to provide fire protection. Further protection is possible if the void between the cladding and the member is filled with a mineral wool insulation and fire protection layer. Blontrock *et al* (2001) report fire tests on **concrete** slabs with externally bonded CFRP. Without protection, the adhesive became ineffective at 47–69°C. The fire resistance (before loss of structural strength) of the strengthened slabs protected with cladding was at least as great as for the unstrengthened, unclad concrete slabs.

6.6 DETAILING

6.6.1 The width of preformed strengthening

Ideally a single piece of preformed FRP would be applied across the whole width of a member. As the plate width (b_f in Figure 6.3) increases, however, it becomes more difficult to expel all of the air from within the adhesive layer and achieve the desired 100 per cent bond cover between the mating surfaces. Many pultruded strengthening plates come in widths of 150–200 mm, while prepreg sheets may be up to about 300 mm wide.

Where multiple strips of strengthening material are applied across the width of a member, a minimum separation should be specified between the plates (s in Figure 6.3), to allow air and excess adhesive to escape.

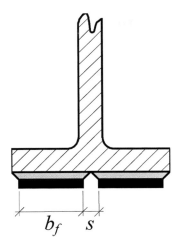

Figure 6.3 *Limits on the geometry of the strengthened section*

6.6.2   Detailing to reduce the stress concentration at the end of a plate

The elastic adhesive joint analysis in Section 5.3.1 shows that the stress concentration in the adhesive at the end of a strengthening plate can be reduced by changing some parameters in Equations 5.47 and 5.57. The material properties of the FRP and the substrate are fixed, but local alterations can be made to the geometry of the strengthening material, the geometry of the adhesive layer and the stiffness of the adhesive.

Perhaps the simplest method of reducing the adhesion stresses is to reduce the thickness of the FRP, t_f, while increasing its breadth, so as to keep the cross-sectional area unchanged. This is feasible only when there is a sufficient usable surface area on the existing structure to accommodate the optimum width of FRP.

Several other options are available for reducing the adhesive stresses near the end of the FRP, which are shown in Figure 6.4. All of these methods are used in the aerospace industry, where FRP patches are used to repair metallic substrates (Baker, 1996).

1. A spew fillet of adhesive beyond the end of the plate helps to reduce the local stress and also provides some protection against environmental attack of the adhesive layer.

2. The stress concentrations can be greatly reduced by tapering the end of the FRP plate, either gradually or in a series of steps. This is practical only when the thickness or width of the plate is sufficient. In the case of preformed plates, end tapers require special manufacture or fabrication, resulting in significant additional cost compared with standard plates of uniform section.

3. Accompanying the FRP taper by a matching counter taper in the adhesive thickness further reduces local adhesive stresses.

4. Reducing the stiffness of the adhesive towards the end of the plate reduces the stress concentration, but may be impractical and unreliable.

5. Mechanical restraint can be used to restrain the FRP after a debonding failure. Bolted clamps can be used to clamp the strengthening material against the metallic member. Additional FRP can be wrapped around the strengthened flange of a beam to provide the necessary restraint. A further advantage of mechanical restraint is that confinement of the adhesive improves its resistance. Ideally, the bolts should not pass through the FRP, as this will severely reduce the plate's strength.

The designer should consider the sensitivity of tapered strengthening material to regions of unbonded adhesive. For example, tapering the FRP plate (option 2) reduces the stress concentration at the end of the plate. However, should a crack form in the adhesive at the end of the plate the crack may be unstable, leading to sudden, catastrophic failure. As the crack length increases, the thickness of the FRP plate at the crack tip becomes greater and consequently the stress concentration increases, further driving propagation of the crack.

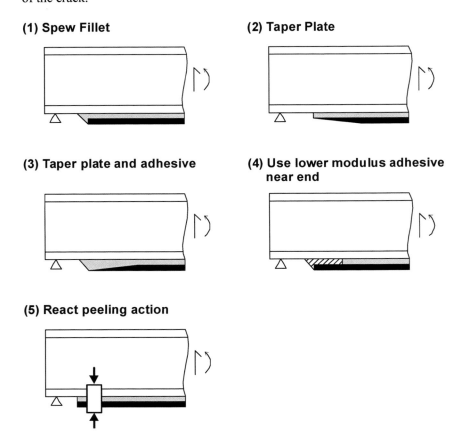

(1) Spew Fillet

(2) Taper Plate

(3) Taper plate and adhesive

(4) Use lower modulus adhesive near end

(5) React peeling action

Figure 6.4   *Methods of reducing the stress concentration in the adhesive at the end of a plate*

Gradual anchorage for prestressed FRP strengthening systems

Stöcklin and Meier (2001) have developed a method of gradual anchorage for prestressed FRP. It allows prestressed FRP to be applied without the need for mechanical anchorage. Their procedure has been successfully applied to full-size concrete beams in laboratory conditions, and is shown schematically in Figure 6.5.

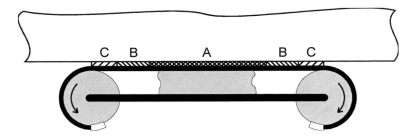

Figure 6.5   *Gradual anchorage of prestressed strengthening material*

Gradual anchorage is generated by a stepwise sequence of load transfer and bonding, as shown schematically in Figure 6.5. The FRP is stressed, and bonded to the structure over a part of its length (A). Once the adhesive has partially cured, a proportion of the prestress is transferred to the adhesive bond by reducing the prestress load at the end of the plate. A further length of bond is then formed (B). The process is repeated for further lengths of bond (C), until all of the prestress has been transferred. The stressing rig is temporary and is removed once the application procedure has been completed.

6.6.3 Curtailment of strengthening material

The ends of a piece of FRP strengthening can be curtailed to suit the bending moment distribution along a member. However, the strengthening should not be curtailed at the theoretical section at which a sectional analysis indicates that it is no longer required. It should be extended by an "anchorage length" c beyond the theoretical curtailment point.

No research appears to have been carried out to determine the anchorage length, c, required for externally bonded FRP on a metallic substrate. Pending further research work, the following guidelines are suggested, based upon both theoretical principles and current curtailment practice in analogous areas. The designer should however be satisfied by testing or analysis that the anchorage length is satisfactory.

The anchorage length c is built up from three separate components:

$$c = c_1 + c_2 + c_3 \qquad (6.17)$$

where:

c_1 *The length required for stress development in the laminate.*
An elastic adhesive joint analysis can be used to determine the shear and peel stress distributions near the end of the FRP and thus the length of joint required to develop the stress in the plate. For example, Figure 5.9 was constructed using equations A5.19 and A5.33 in Appendix 5.

c_2 *Variability in the position at which the strengthening material is no longer needed*
BS 8110:1997 recommends that steel reinforcement in concrete is extended by at least the effective depth of the beam beyond the section at which it is required. c_2 should similarly be related to the depth of the beam section.

c_3 *Delamination damage to the adhesive near the end of the laminate*
Delamination damage could occur due to overload, fatigue or environmental attack. In the absence of other information, c_3 should be taken as 50 per cent of the stress development length (c_1), or the width of the piece of strengthening material, whichever is most critical.

An anchorage length of 72 t_f (where t_f is the plate thickness, for $b/t \geq 40$) can be deduced for steel plates, from the information in BA 30/94 (Department of Transport, 1994). This should be adopted as a minimum for FRP strengthening plates.

The cross-sectional area of the FRP can be reduced along a beam in accordance with the bending moment envelope. This can be achieved by applying multiple plates (and staggering the ends of the plates), or by prefabricating (in which case the number of laminates is varied along its length). The adhesion strength should be checked at each plate end.

Peel stresses due to changes in direction of the strengthening material

The surface to which the strengthening material is applied should ideally be flat, but in reality will contain imperfections. When the surface is curved or kinked, it is important to ensure that the proposed form of strengthening can conform to the profile and that the peel stresses associated with the curvature of the FRP are acceptable.

In some cases, it may be desirable to use a wet lay-up method, allowing the FRP to conform to the curved profile of a member without being pre-strained. As indicated in Figure 6.6, a concave change in direction of the strengthening material results in tensile peel stresses being induced through the adhesive, tending to separate the FRP from the substrate.

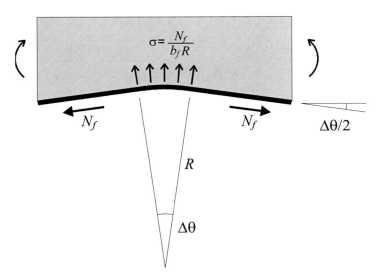

Figure 6.6 *Peeling stresses due to changes in the direction of the strengthening material*

The peel stress due to a change in direction can be determined from the force in the FRP, N_f, and the radius of curvature, R:

$$\sigma = \frac{N_f}{b_f R} \tag{6.18}$$

where b_f is the width of the FRP and R is given by:

$$\frac{1}{R} = \frac{\Delta\theta}{\ell} \tag{6.19}$$

$\Delta\theta$ being the angle turned through and ℓ the length of the circular transition.

The peel stress can be reduced by increasing the radius of curvature (R) of the change in direction. Avoid abrupt changes in direction of the strengthening material. Any change in direction should be made as gentle as possible. In practice, it will be necessary to:

- grind off local protrusions should be wherever possible
- fill local surface depressions with a layer of adhesive putty (Section 7.3.2)
- use a tapering filler adhesive layer to provide a smooth transition and provide an adequate radius of curvature R to control the peeling stress (Figure 6.7) where a change in direction of the FRP is unavoidable.

In extreme cases, the peel stresses should be carried using a transverse strap or mechanical anchorage. For example, an FRP strap can be bonded beneath the surface of the FRP strengthening material, and bonded onto the existing structure to either side.

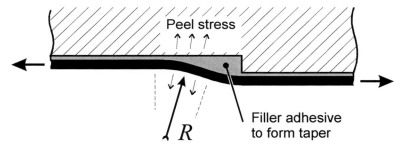

Figure 6.7   *Producing a smooth change in direction using filler adhesive*

The application of strengthening material in multiple directions

When strengthening needs to be applied in two or more directions, the strips of strengthening material will have to cross. Just as with surface irregularities, filler adhesive should be applied to produce a smooth profile for the laminates and control peel stresses in the adhesive. The increased thickness of adhesive should be taken into account in the analysis of the adhesive stresses (Section 5.3.1).

Strengthening at connections

If continuity of the strengthening material is required across an existing connection detail, the FRP must not induce unacceptable tensile peel stresses in the adhesive.

The metallic member can be built up locally using a tapered filler piece, permitting continuity of the strengthening material over the connection. Alternatively, the strengthening material can be overwrapped with wet lay-up transverse FRP straps to carry the normal actions.

6.6.5   **Strengthening bolted or riveted structures**

In built-up wrought iron or early steel members, rivets or bolts joining the component parts obstruct the surface to which FRP strengthening is to be applied. One of the following strengthening options could be adopted.

1.   The FRP can be placed in strips between the rows of rivet heads (Figure 6.8, left-hand side).

2.   A preformed plate with holes to accommodate the rivet heads can be bonded to the surface of the plate.

3.   FRP plates can be applied to a flat surface created by a make-up layer covering the rivet heads (Figure 6.8, right-hand side).

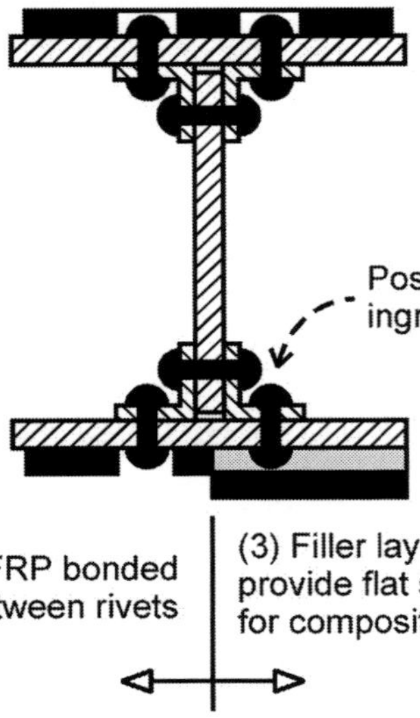

(2) Holes in FRP to accommodate rivets

Possible water ingress

(1) FRP bonded between rivets

(3) Filler layer to provide flat surface for composite

Figure 6.8 *Strengthening a fabricated member with rivets*

Option 1 is the most straightforward and is recommended where possible. The holes in the second approach create stress concentrations in the FRP, which tend to negate any benefit obtained from the greater width of strengthening laminate. The make-up layer in the third option should be checked for creep. For example, a thick adhesive layer will be unsuitable, but a GFRP plate could be used.

A concern with Options 2 and 3 is that water could find its way along the shaft of the rivet (as indicated in the figure) and then into the adhesive joint between the FRP and the substrate. This would cause both degradation of the adhesive and corrosion of the metallic substrate, potentially leading to failure of the adhesive bond in the long term. Protection against water ingress can be provided by painting with an epoxy-based coating, although this requires rigorous inspection and maintenance, and should be recorded in the health and safety file and six-monthly inspection protocols.

Strip strengthening material (1) does not enclose the bottom of the rivet. Bassetti (2001) and Bassetti *et al* (2000a, 2000b) applied narrow prestressed FRP strips to riveted, fabricated steel girders taken from a truss railway bridge. He showed that this is an effective method of extending the fatigue life of a riveted structure, since the prestress acts to close fatigue cracks, which typically initiate adjacent to rivets (Section 6.4.3).

6.6.6 Co-ordination with finishes

A wide variety of coatings or cladding can be applied over the FRP strengthening material (although, as discussed in Section 7.7, these make inspection more difficult). Before a coating is applied, it should be checked for chemical compatibility with the strengthening material.

6.6.7 Details necessitated by the electrical conductivity of carbon fibres

Carbon fibres (unlike aramid or glass fibres) are good electrical conductors. CFRP strengthening systems must therefore address the following potential issues:

* galvanic interaction between the carbon fibres and the metallic substrate
* currents induced in the carbon fibres by nearby electrical installations
* lightning strike and flashovers.

Galvanic corrosion

When graphite, the form of carbon used for strengthening, comes into direct contact with a metal, a current flows between them, causing the metal to corrode. The following three conditions are necessary for galvanic corrosion.

1. Two different conductive materials must be present.

2. An electrical connection joining the different materials must exist.

3. An electrolyte must be present.

The electrolyte will usually be surface moisture on a structure, such as rainwater or condensation, which has absorbed pollutants, sea salt or processing residues from the metal.

The corrosion potential between different metals and conducting non-metals is ranked by the electropotential series. Carbon is widely separated from steel and iron on the electropotential series, making galvanic corrosion likely. Galvanic coupling can also lead to corrosion of normally passive metals, such as stainless steel or titanium by raising the potential of the metal (Tucker *et al*, 1990).

The aerospace industry has been successfully joining aluminium and carbon fibres (which are more widely spaced on the electropotential series than steel and carbon) for a number of years. Suitable design and detailing is used to avoid adverse effects.

Galvanic corrosion can be prevented by interposing an insulating layer in the galvanic circuit. There are three possibilities.

1. Cover the entire surface of the metal to which the CFRP is not applied (for example, by painting the existing structure).

2. Cover the surface of the CFRP.

3. Interpose an insulating layer between the CFRP and metal.

The first approach usually involves painting a large area of the metal. Damage to the protective coating (such as a stone chip) leads to rapid localised corrosion. The CFRP will usually include a surface layer of resin, which acts as an insulator, but this is also prone to damage, especially at the cut ends of a plate. The most durable and practical solution is to provide an insulating layer between the CFRP and the metal.

Long-term galvanic corrosion tests at the Swiss Federal Laboratories for Materials Testing and Research (EMPA) rely on the adhesive as the insulating layer between the CFRP and aluminium. After about 10 years of exposure their condition is reported to be very good. The thickness and continuity of the insulating adhesive layer, however, is difficult to guarantee, especially with historic structures where the surface of the metal is not completely flat.

It is therefore recommended that a glass fibre (or other insulating fibre) layer be interposed between the metal and the CFRP (an approach used for several years in the aerospace industry). This guarantees a minimum separation between the carbon fibres and the substrate, and the glass fibres and resin or adhesive act as an insulating layer. Insulation is also required at any mechanical anchorages, such as bolts or clamps, to avoid a galvanic circuit.

The glass layer can be incorporated in preformed strengthening plates (with the advantage of high quality assurance) or can be introduced separately by wet lay-up. The minimum recommended thickness of the insulating layer is 0.5 mm.

West (2001) has carried out a thorough series of tests, showing that a GFRP insulating layer is an effective method of preventing galvanic corrosion between steel and CFRP in a bridge-strengthening project. West also showed that a fillet of adhesive is important around the edge of the FRP to act as a sealant and thus prevent corrosion due to the ingress of water into the joint.

Induced currents

If externally bonded CFRP is installed near high-voltage alternating electrical installations, induced currents are generated in the carbon fibres. Induced currents have been used in a controlled manner to heat the CFRP and accelerate curing of the adhesive. It is possible that uncontrolled induced currents would heat the adhesive above its glass transition temperature T_g, leading to damage or failure of the strengthening system.

Metal structures near high-voltage installations (such as railway bridges above traction supply lines) are connected to the power return. It is not possible to connect the carbon fibres, since they are numerous and embedded in an insulating resin. A connection to the steel substrate would also promote galvanic corrosion.

Calculations of induced currents within carbon fibre composites suggest that this effect is not significant. For a carbon fibre plate 8 mm thick, 200 mm long bonded to a steel plate 6 mm thick, 200 mm long located 300 mm from a 25 kV AC high-voltage conductor (300 Amps) the calculated (induced) power dissipated within the carbon fibre plate is 200 μW.

Indicative calculations suggest that induced currents are not likely to be a problem, but CFRP strengthening is not currently recommended for use near overhead power lines. Two methods to avoid induced current problems are to use a non-conducting strengthening material such as AFRP or GFRP, or to enclose the CFRP in a Faraday cage (a metal mesh), taking care to avoid galvanic effects. Note that AFRP strengthening must be protected from excess humidity, as moisture absorption allows currents to be induced in aramid fibres.

Lightning strike and flashovers

CFRP is frequently bonded to the underside of a structure, where it is unlikely to be exposed to the risk of lightning strike. If CFRP is applied to the upper surface of a metallic beam, a lightning strike in close proximity to the CFRP could generate sufficient heat to damage the adhesive interface and hence the strengthening material. Similar damage could be caused by a flashover between overhead power supplies and CFRP strengthening.

Where CFRP is applied in a place exposed to lightning strike or flashover, an alternative path should be provided. For example, a lightning conductor should be provided to attract the lightning, or a metal grid used to enclose and thus protect the strengthening material.

6.6.8   Details to minimise the effects of accidental damage

FRP strengthening material may be accidentally damaged, either due to a lack of familiarity with the material, or due to impact damage.

As discussed in Section 7.7, it is preferable to leave the FRP strengthening material exposed and not covered by paints or other finishes, so that it is visible during future maintenance and modifications to the structure. Furthermore, consideration should be given to attaching warning labels on, or adjacent, to the strengthening system to highlight that the FRP must not be damaged. TR57 (Concrete Society, 2003) shows example warning labels).

FRP strengthening is particularly susceptible to impact damage when it is exposed on the soffit of a low road over-bridge, as it is vulnerable to damage in the event of a bridge strike by an over-height vehicle. Details to protect the strengthening material should be specified. For example, non-structural sacrificial material thicker than the strengthening can be added to the soffit of the bridge on either side of the FRP strengthening. It will be damaged before the FRP, which will provide a degree of protection for the FRP.

In some situations it may be feasible to locate the strengthening on the top surface of the bottom flange, thereby protecting it from impacts from below. As an added benefit it would not encroach into the headroom.

6.7   BIBLIOGRAPHY

ASCE (1984). *Structural plastics design manual*. ASCE Manuals & Reports on Engineering Practice No 63, American Society of Civil Engineers, Reston, VA

Aylor, D M (1993). "The effect of a seawater environment on the galvanic corrosion behavior of graphite/epoxy composite coupled to metals". In: C E Harris and T S Gates (eds), *High temperature and environmental effects on polymeric composites*. STP 1174, ASTM, West Conshohocken, PA

Baker, A (1996). "Joining and repair of aircraft composite structures". *Mech engg trans*, vol ME21, nos 1 and 2

Bank, L C and Mosallam, A S (1992). "Creep and fatigue of a full-size fibre-reinforced plastic pultruded frame". *Comp engg*, vol 2, no 3, pp 213–227

Barnes, R A and Mays, G C (2001). "The effect of traffic vibration on adhesive curing during installation of bonded external reinforcement". *Proc Inst Civ Engrs, Structures and buildings*, vol 136, no 4, pp 403–410

Bassetti, A (2001). "Lamelles précontraintes en fibres carbone pour le renforcement de ponts rivetés endommagés par fatigue". PhD thesis EPFL 2440, École Polytechnique Fédérale, Lausanne

Bassetti, A, Nussbaumer, A and Colombi, P (2000a). "Repair of riveted bridge members damaged by fatigue using CFRP materials". In: G Pascale (ed), *Proc conf advanced FRP materials for civil structures, Bologna*, pp 33–42

Bassetti, A, Nussbaumer, A and Hirt, M (2000b). "Crack repair and fatigue life extension of riveted bridge members using composite materials". In: A-H Hosny (ed), *Proc bridge engg conf 2000, ESE-IABSE-FIB, Sharm El-Sheikh*. Egyptian Soc Engrs,

Cairo, vol I,
pp 227–238

Bellucci, F (1992). "Galvanic corrosion between nonmetallic composites and metals II. Effect of area ratio and environmental degradation". *Corrosion*, vol 48, pp 281–291

Blontrock, H, Taerwe, L and Vandevelde, P (2001). "Fire testing of concrete slabs strengthened with fibre composite laminates". In: *Proc 5th int conf FRP reinf conc strucs (FRPRCS-5), 16–18 Jul, Cambridge*. Thomas Telford, London, pp 547–556

Boyd, J *et al* (1991). "Galvanic corrosion effects on carbon fiber composites". *Proc 36th int SAMPE symp, 14–18 Apr, San Diego*. SAMPE, Covina, CA

CIRIA (1977). *Rationalisation of safety and serviceability factors in structural codes.* Report 63, CIRIA, London

Clarke, J L, ed (1996). *Structural design of polymer composites. Eurocomp design code and handbook.* E & FN Spon, London

Concrete Society (2000). *Design guidance for strengthening concrete structures using fibre composite materials.* Technical Report 55, Concrete Society, Crowthorne

Concrete Society (2003). *Strengthening concrete structures with fibre composite materials: acceptance, inspection and monitoring.* Technical Report 57, Concrete Society, Crowthorne

Denton, S R (2001). "Analysis of stresses developed in FRP plated beams due to thermal effects". In: J G Teng (ed), *Proc int conf FRP composites in civ engg (CICE 2001), 12–15 Dec, Hong Kong*, pp 527–536

Department of Transport (1994). *Strengthening of concrete highway structures using externally bonded plates.* BA 30/94 (DMRB vol 3, sec 3, pt 1), HMSO, London

Findley, W N (1971). "Combined stress creep of non-linear viscoelastic material". In: A L Smith and A M Nicolson (eds), *Advances in creep design*. Applied Science Publications, London

Griffith, W I, Morris, D H and Brinson, H F (1978). *The accelerated characterisation of composite materials.* VPI engineering series VPI-E-78-3. Virginia Polytechnic Institute College of Engineering, Blacksburg, VA

Hambly, E C (1991). *Bridge deck behaviour.* 2nd edn, Spon Press, London

Hart-Smith, L J (1994). "The key to designing durable adhesively bonded joints". *Composites*, vol 25, no 9, pp 895–898

Highways Agency (1996). *The assessment of steel highway bridges and structures.* BD 56/96 (DMRB vol 3, sec 4, pt 11), Stationery Office, London

Highways Agency (2001a). *The assessment of highway bridges and structures.* BD 21/01 (DMRB vol 3, sec 4, pt 3), Stationery Office, London

Highways Agency (2001b). *Loads for highway bridges.* BD 37/01 (DMRB vol 1, sec 3, pt 14), Stationery Office, London

Hollaway, L C and Head, P R (2001). *Advanced polymer composites and polymers in the civil infrastructure*. Elsevier Science

Jones, C (1994). "Fatigue of GFRP". *J Composite materials*, vol 28, pp 309–327

Mertz, D R, Gillespie, J W, Chajes, M J and Sabol S A, (2001). *The rehabilitation of steel bridge girders using advanced composite materials*. IDEA program final report, contract no NCHRP-98-ID051, Transportation Research Board, National Research Council

Miller, T C, Chajes, M J, Mertz, D R and Hastings, J N (2001). "Strengthening of a steel bridge girder using CFRP plates". In: *Proc New York City bridge conf, 29–30 Oct, New York*

Moy, S S J (2001). "Early age curing under cyclic loading – a further investigation into stiffness development in carbon fibre reinforced steel beams". Link & Surface Transport Programme, Dept of Civil & Environmental Engg, Univ Southampton

Moy, S S J (2002). "Early age curing under cyclic loading – an investigation into stiffness development in carbon fibre reinforced steel beams". In: *Proc ACIC 2002, inaug int conf use of advanced composites in construction, 15–17 Apr, Univ Southampton*. Thomas Telford, London

Rajagopalan, G, Immordino, K M and Gillespie, J W (1996). "Adhesive selection methodology for rehabilitation of steel bridges with composite materials". In: *Proc 11th tech conf on composites*. American Society for Composites, Atlanta, p 222

Stöcklin, I and Meier, U (2001). "Strengthening of concrete structures with prestressed and gradually anchored CFRP strips". In: *Proc 5th int conf FRP reinf conc strucs (FRPRCS-5), 16–18 Jul, Cambridge*. Thomas Telford, London, pp 291–296

Tucker, W C, Brown, R and Russell, L (1990), "Corrosion between a graphite/polymer composite and metals". *J composite materials*, vol 24, no 1, pp 92–102

West, T D (2001). *Enhancement to the bond between advanced composite materials and steel for bridge rehabilitation*. CCM Report 2001-04, University of Delaware Center for Composite Materials

British Standards

BS 476-11 to 20:1987. *Fire tests on building materials and structures. Method for determination of the fire resistance of elements of construction*

BS 5400-1:1988. *Steel, concrete and composite bridges. General statement*

BS 5400-10:1980. *Steel, concrete and composite bridges. Code of practice for fatigue*

BS 5950:2000. *Structural use of steelwork in building. Code of practice for design. Rolled and welded sections*

BS 7608:1993. *Code of practice for fatigue design and assessment of steel structures*

BS 8110-1:1997. *Structural use of concrete: Part 1: Code of practice for design and construction*

BS EN 1990:2002. *Eurocode. Basis of structural design*

BS EN 2378:1994. *Fibre reinforced plastics. Determination of water absorption by immersion*

7 Installation and quality control

High-quality workmanship is essential with externally bonded FRP strengthening systems. Inadequate installation by untrained personnel is likely to result in poor performance and premature failure of the strengthening works. This chapter describes installation of an FRP strengthening scheme.

Section 7.1 outlines aspects of a strengthening scheme that are essential to a high quality installation, including test procedures. An appropriate working environment is discussed in Section 7.2, including the operational requirements of the installation, the safety of personnel, and the surrounding environment.

Installation processes for the FRP strengthening are described in the subsequent sections. Section 7.3 describes preparation of the metallic structure prior to strengthening. Sections 7.4 and 7.5 cover the application process for the strengthening material, and the curing of the resin/adhesive. Permanent load transfer methods (using prestressed FRP, or load-relief jacking) are described in Section 7.6, while Section 7.7 covers the application of finishes over the strengthening.

Much of the chapter is common to all methods of strengthening, but specific requirements are identified where necessary.

7.1 Quality control

Failure of an FRP-strengthened structure is most evident as large-scale delamination of the strengthening. Failure is not always so obvious, however. If post-installation inspection (see Section 8.1) reveals unacceptable defects in the adhesive joint, the only recourse is likely to involve removing the defective installation and replacing it with new FRP strengthening material. It is sensible to use appropriate management procedures to minimise the potential for installation errors and defects both before and during the installation process. It is essential that the workforce carrying out the installation work are familiar with the requirements of FRP strengthening systems, which include:

- correct storage and handling of the *materials* involved (Section 7.2.4)
- the importance of surface preparation and *cleanliness* (Section 7.3.2)
- the *requirements* of the installation process (Section 7.4)
- *why* the materials should be installed this way (the intent of the strengthening scheme)
- the *implications* of an incorrect installation.

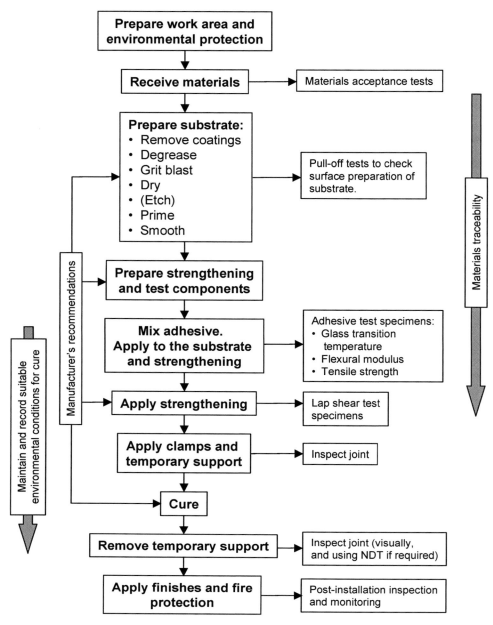

Figure 7.1 *Strengthening with preformed plates – the installation process and quality procedures*

The quality of the installation also depends upon the quality and practicality of the design of the strengthening works. Designs that fail to take into account construction tolerances or that fail to minimise difficulties or risks in installation may contribute to poor performance of the finished works. Details such as the orientation and position of fibres are more easily controlled under factory conditions than by *in-situ* methods. By using pultruded or prepreg composites to strengthen a structure, a potential source of error can be removed from the installation process.

It is recommended that a specialist contractor with previous experience of both FRP strengthening systems and metallic structures be employed to carry out the strengthening work. This is discussed further in Section 9.1. The contractor should submit a method statement before starting strengthening works.

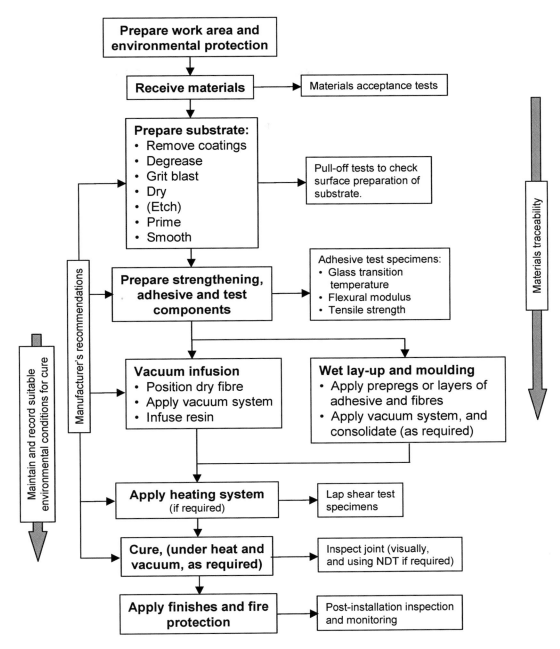

Figure 7.2 In-situ *strengthening techniques – the installation process and quality procedures*

The designer of the strengthening scheme should have an in-depth understanding of strengthening works and the importance of correct installation. The requirements of the strengthening works should be clearly documented in a detailed materials and workmanship specification. To ensure that salient aspects of this knowledge are transferred to site and fully understood, **it is recommended that the designer be represented during the works to ensure that the scheme is being implemented correctly and to provide advice on any significant variations required to the process during the works.**

Specific quality procedures are described in the sections that follow. The installation procedure is summarised in Figures 7.1 and 7.2, along with the tests required.

7.1.1 Test procedures

Tests characterise the quality of various elements of the structure, namely:

- the composite strengthening material
- the adhesive
- bond of the strengthening material to the substrate (including the correct surface preparation of the composite and substrate).

Tests are required at three principal stages of a structural strengthening scheme.

1. *Characterisation tests*
   The properties of the strengthening materials need to be known before design. The manufacturer usually conducts these tests.

2. *Quality control (proof) tests*
   The properties of the supplied strengthening material must be checked before installation. There is no need to test materials before installation if the supplier has an approved quality system, if the supplier's products are sample-tested by an independent testing centre and if the supplier provides certificates of conformity. If the supplier does not have an approved quality system, material samples must be sent to an independent testing centre for proof-testing.

   Further quality control tests (Section 7.4.5) are required to assess whether the strengthening scheme has been properly installed. Specimens should be prepared at the time of installation for later testing. It is of benefit to test some of the specimens before the strengthening operation has been completed, so that faults can be rectified at the earliest opportunity.

3. *Inspection and monitoring tests*
   It is advisable to prepare additional specimens at the time of installation, to be set aside for future testing. These allow the long-term performance of the strengthening system to be assessed, without requiring potentially damaging tests on the strengthening material itself.

   Test samples can be prepared for each of the quality control test methods listed below, including samples of FRP bonded to the metallic member that are suitable for pull-off tests. The specimens should be chosen to provide information on the critical components of a strengthening system. No fewer than three of each test sample should be available on each future occasion that monitoring tests are anticipated. TR57 (Concrete Society, 2003) gives advice on such test samples.

Table 7.1 lists British and international test standards. Equivalent ASTM standards could also be used. The test method specification should be recorded and should not be changed part way through a project. For example, tests carried out by the adhesive manufacturer should be undertaken to the same specification as tests carried out on specimens produced on site. Where a test specimen is prepared on site, the preparation method and curing conditions should be as close as possible to those of the strengthening works.

Composite material

The acceptance tests for composite strengthening material should include the tensile stiffness and strength.

Tensile tests can be problematic for thick preformed strengthening material, because of the limitations of the testing machines available. Flexural strength and stiffness tests give conservative results, due to the laminated nature of the composite. Laminate theory should be used to predict the in-plane tensile properties of the plates from flexural tests, the results of which should be checked by comparing in-plane tensile properties determined by tensile and flexural tests for thinner laminates.

Adhesive

Tests to characterise the adhesive include lap-shear tests and tests to determine the glass transition temperature. The material supplier will normally provide the glass transition temperature, which will have been evaluated by one of two test methods: differential scanning calorimetry (DSC) or dynamic mechanical thermal analysis (DMTA). Although either method is acceptable, they give slightly different values for the glass transition temperature, so it is essential that the tests carried out during installation use an agreed standard method – normally that used by the material supplier.

Table 7.1 *Materials testing standards*

Composite

BS EN ISO 527:1997	In-plane tensile properties.
	(527-4 for isotropic and orthotropic FRPs; 527-5 for unidirectional FRPs)
BS EN ISO 14125:1998	Flexural properties
BS EN ISO 14126:1999	In-plane compressive properties
BS ISO 15310:1999, BS EN ISO 14129:1998	In-plane shear properties
BS EN ISO 14130:1998	Interlaminar shear strength in multiple-ply laminates
BS ISO 1268:2001	Methods of producing test plates
BS ISO 11359-2:1999	Coefficient of thermal expansion

Adhesive

BS EN 14002:2003	Pot life
BS 2782-3:1976: Methods 320A to 320F	Tensile properties of adhesive
BS 6319-3:1990	Flexural properties of adhesive
BS EN 1770:1998	Coefficient of thermal expansion
BS 5350-C5:2002	Lap-shear tests for bond strength (single or double overlap)
BS ISO 11357-2:1999	Glass transition temperature by differential scanning calorimetry (DSC)
ISO 6721-1:2001/ ASTM E 1640-99	Glass transition temperature by dynamic mechanical thermal analysis (DMTA)
BS EN ISO 899-1:2003 and	Creep behaviour – tensile creep
BS EN ISO 899-2:2003	Creep behaviour – flexural creep
BS 6319-8:1984	Moisture resistance

Adhesive bond

BS 5350-C5:2002	Lap-shear tests for bond strength (single or double overlap)
BS 1881-207:1992	Pull-off test (for concrete)
BS EN ISO 4624:2003 / BS 3900-E10:2003	Pull-off test (for paints and varnishes)

Lap-shear test for adhesive bond

A *lap-shear test* is a consistent and reliable method for assessing the adhesive bond between two surfaces. Lap-shear tests are conducted at two stages during the strengthening process.

1. To determine the strength of the adhesive joint, for use in design.

2. To check that the required bond strength is being obtained during installation.

Either a single lap or a double lap shear specimen can be used, as shown in Figure 5.11. Single-sided lap-shear tests using thin laminates should be avoided, as the load is applied eccentrically and these tests do not model the inflexible metallic substrate correctly.

The lap-shear tests must characterise failure of the weakest component of the adhesive joint. If the surface preparation has been carried out correctly, the weakest component is usually the composite. It is recommended that the following lap-shear tests be undertaken.

1. To provide design values:
    *   double lap-shear tests between composite and metal to determine the peak strength of the adhesive joint (for the metal present in the structure to be strengthened)
    *   single composite-composite lap-shear tests to provide a baseline for lap-shear tests during installation and further information on the strength of the adhesive bond (assuming that the composite is the weakest component).

2. During installation:
    *   single composite-composite lap-shear tests.

Lap shear specimens must receive the same surface preparation as in the final works (Section 7.3), and have the same thickness of adhesive. It is also important that they are cured under the same temperature and pressure regime as in the installation of the works and that they are then conditioned under a realistic hygrothermal environment.

The average lap-shear stress at failure reported by adhesive suppliers is usually in the range 14–22 MPa. It is the peak adhesive strength that is of importance in design, however. The lap-shear test results are back-analysed to find the peak adhesive strength (Section 5.3.1), allowing the peak adhesive stress to be found irrespective of the geometry or substrates used. To calculate the peak stress at failure, or to compare the adhesive from different manufacturers, it is important to record the geometry, curing method and environmental conditioning of the test specimen.

Pull-off test for adhesive bond

For externally bonded FRP strengthening material applied to metallic structures, a test is required to check that sufficient surface preparation has been carried out to give a good adhesive bond (Section 7.3.2). A pull-off test can be adopted, using the adhesive that will be used to strengthen the structure, under the same environmental and application conditions.

The purpose of the tests is to:

(a) indicate the quality of surface preparation

(b) verify that the adhesive is compatible with the prepared surface.

Two standards are listed in Table 7.1, both of which describe the same pull-off test method. The first is intended for assessing the strength of concrete, while the second tests adhesion of paint or varnish to the substrate. As the adhesive interface is stronger than either paint or concrete, the use of a smaller, 20 mm-diameter "dolly" is recommended.

7.1.2   Record keeping

At all stages during the project, thorough records should be kept to provide evidence of the work carried out and its quality. The structure's health and safety file should be updated with a permanent record of the strengthening work, including:

- the work carried out

- the form of the strengthening scheme, including the design philosophy

- the properties of the materials used, including all test results

- traceability of the materials used

- the conditions under which the work was undertaken

- evidence that the work exceeded or at least reached the specified minimum quality

- any initial defects in the system (such as small areas of delamination)

- future testing, inspection and maintenance requirements

- instructions for the use of any installed monitoring system

- measures to be taken in the event of the structure being damaged by a variety of means, which will vary according to the nature and the extent of the damage. Should the structure need to be closed to allow any repairs necessary to be carried out, this should be indicated.

In addition to written records, it is desirable to keep specimens of the materials used (Section 7.1.1).

7.2   WORKING ENVIRONMENT

7.2.1   Health and safety

As in any construction work, the health and safety implications of the installation must be examined before work begins. Risks have to be established and measures taken to minimise the risk to anybody involved, including the workforce and members of the general public. The contractor must prepare a method statement, which has to be approved by the client before any strengthening work begins.

All work must comply with the statutory procedures applicable to the location of the structure. In the UK, the installation must comply with the Health & Safety at Work etc Act, the Control of Substances Hazardous to Health Regulations (COSHH) and the Construction (Design & Management) Regulations (CDM). Special safety requirements apply in certain industries or locations, such as railways and nuclear installations.

All personnel should be issued with the correct protective equipment (including eye protection, gloves, overalls, ventilators and respirators). The material suppliers will provide information about the hazards associated with each of the products used in the strengthening scheme (the COSHH information), including the strengthening material, the uncured adhesive, and solvents and abrasives used during surface preparation. The following list highlights some potential hazards that should be considered, but the supplier should be consulted for the particular materials being used.

- protection is needed from loose or protruding fibres, which are often sharp and can easily penetrate the skin
- machining and cutting of strengthening material and preparing the substrate creates dust, which should be extracted to protect operatives from inhalation
- adhesives cause skin irritation and sensitisation
- volatile emissions from the adhesives should be extracted and not inhaled
- cleaning solvents such as acetone and propan-2-ol are highly flammable.

During the planning of the work, provision should be made for a safe means of access to the work site, and a safe working environment within the work site while the installation is carried out (Section 7.2.3).

When compared with other strengthening techniques, such as steel plate strengthening, FRP has health and safety benefits. In particular, composite components are lightweight and require handling for shorter periods of time. (Large, thick composite plates may still be sufficiently heavy to rule out manual lifting.) Less plant and access equipment will need to be present on site, with further safety benefits.

7.2.2   Environmental issues

In addition to the safety of personnel, potential damage to the surrounding environment should be assessed.

The main practical environmental concern is the need to contain, collect and properly dispose of all products and by-products of the installation. These products include:

- the strengthening materials
- abrasives used in grit blasting or mechanical abrasion
- paint layers removed during surface preparation. These can include lead, which must not be inhaled

- metallic material removed during surface preparation. The arisings from grit-blasting cast iron are very active and environmentally damaging

- solvents used during degreasing

- waste adhesive (including adhesive that has exceeded its pot life, and adhesive wiped off after the bonding operation)

- waste composite (including peel plies)

- emissions from the adhesive during curing (adhesives provided by the manufacturers generally have low volatile emissions).

Where work is being undertaken above a watercourse, particular care will be necessary to prevent noxious substances causing contamination.

The lightweight nature of composite components reduces transportation requirements, both in terms of the materials to be delivered, and other equipment such as heavy lifting equipment that will not be required on site.

7.2.3 Work space

The quality of the strengthening works can be greatly improved by providing a controlled environment in which the strengthening work can be undertaken. A controlled environment has a number of benefits:

- protection from the weather and airborne contaminants (such as dust or leaves)

- provision of the environmental conditions necessary for adhesive cure (temperature and humidity)

- containment of waste materials, preventing contamination of the surrounding environment

- improved working conditions for personnel. (Protection will be required from dust and fumes contained within the working environment, for example through the use of a suitable extraction system.)

When site constraints (such as possession time) permit, the whole of the zone of the structure to be strengthened should be tented. Otherwise, local enclosures can be formed around the components being strengthened. The enclosures can be heated, to provide the correct temperature for cure of the adhesive and to help prevent condensation collecting on the surfaces to be strengthened (Section 7.5). Figure 7.3 shows an example of tenting and heating arrangements for a cast iron bridge.

Figure 7.3 *Tenting and heating a cast iron bridge during the strengthening operation*

The workspace should be laid out to facilitate the installation process. For example:

- safe, secure storage should be provided for the strengthening materials
- adequate space in which to apply the adhesive to the FRP should be provided
- there must be a clear path from the place where the adhesive is applied to the structure
- the member being strengthened must be easily accessible by operatives carrying the strengthening materials.

The workspace should include a bench with tools for cutting the strengthening materials (if they are not supplied to the correct size); scales and containers for weighing out resin (if not supplied pre-batched); and containers for mixing adhesive (which must either be cleaned before reuse, or disposed of after they have been used for a single batch of adhesive).

7.2.4 Storage and handling of strengthening materials

Provision must be made for correct storage and handling of the strengthening system materials.

The strengthening materials should be stored in a controlled, stable environment, which is dry, away from direct sunlight and dust-free. Uncured adhesives should typically be stored at a temperature of 10–20°C (and must in no circumstances be allowed to freeze); the supplier will be able to advise on the conditions required for a specific product. Pre-impregnated materials will usually require refrigerated storage (-18°C) to arrest curing of the partially cured resin.

Solvent materials are highly flammable, and should be securely stored in non-flammable containers kept in appropriate storage lockers away from potential sources of ignition.

Records should be kept to ensure that specific materials can be traced from delivery to installation. The dates on which the materials are delivered to site and the dates by which they must be used should be recorded and any adhesives that exceed their shelf life must be disposed of. FRP fibres should be stored on site before use, so that they are conditioned to the same temperature as the structure.

Pultruded strips are often supplied on coiled spools. These spools must be of large diameter to avoid permanent curvature of the FRP, which could lead to damage and imperfect bond during installation. Particular care is required while handling UHM CFRP, as the fibres are very brittle.

7.3 PREPARATORY WORK

7.3.1 The adhesive bond

The adhesive bond between the metallic substrate and the FRP is a critical component of the strengthening system. An adhesive joint contains several layers, as shown in Figure 7.4, which are built up during the installation process.

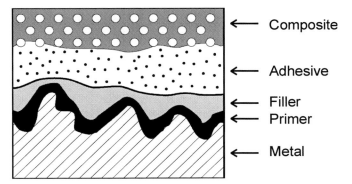

Figure 7.4 *The adhesive joint*

Failure of the joint can be within, or between, any of the layers and will occur at the weakest element. A good adhesive joint requires:

- the absence of any weak layers (such as rust and other metal oxides)
- intimate contact between the adhesive and the adherends
- no interface contamination.

Careful surface preparation is the key to a sound adhesive bond. Weak surface layers must be removed to expose a clean, chemically active surface. Surface contamination should be removed by solvent degreasing.

The integrity of an adhesive bond cannot be assured by inspection alone: a quality bond can only be achieved by careful management of all aspects of the bonding process and strict observation of specified procedures. General recommendations are given in this section, but reference should always be made to the supplier of the material for the particular product being used.

7.3.2 Preparation of the metallic substrate

Surface preparation of the metallic substrate can account for a significant proportion of the cost of strengthening a metallic structure.

Most material suppliers manufacturers can provide recommendations for surface preparation. However, if guidance is not available from the manufacturer, BS 7079:1990 provides guidance for the preparation of steel substrates before painting. This guidance is also relevant for the preparation of metallic surfaces for plate bonding. The surface preparation should be such that a surface equivalent to SA2½ is achieved on the metal. However, note that SA2½ is a visual assessment of the quality of surface preparation and not a measure of the abrasion.

Abrasion

Grit blasting is usually the preferred method of abrasion to expose an active surface prior to bonding. Recirculatory blasting is preferable, to minimise health and safety implications. Wet blasting helps to remove salts from the metallic surface, but the exposed surface must immediately be dried to avoid the formation of corrosion products. Grit-blasting should be carried out in a series of passes, allowing the development of the exposed surface to be monitored. Excessive grit-blasting will damage the existing structure, particularly if it is cast iron.

In industrial or marine atmospheres, rusted surfaces may be contaminated by crystals of ferrous sulphate or chloride, which can be detrimental to the durability of the bonded joint. These crystals can be detected using ferricyanide papers. If grit blasting does not remove the salts, high-pressure water jetting (without detergent) can be used, followed by drying of the exposed surface.

Simple mechanical abrasion using abrasive pads, cloths or wire brushes scores the surface of the metal, tending to produce a folded surface that traps contaminants and moisture, and hence should not be used.

Primer

Corrosion products will form rapidly on the newly exposed metal. It is recommended that the exposed surface is coated as soon as possible, and certainly within two hours of grit blasting. It is normal to apply an adhesive primer coat for this purpose.

A primer/conditioner or adhesive promoter may also be specified by the adhesive supplier to enhance bonding. Primers greatly reduce the variability of the adhesive joint, and the correct choice of primer can improve the stability of the joint with respect to moisture.

Paint primers should not be used under any circumstances, as they do not have sufficient adhesion, and are not compatible with the adhesives used for bonded structural strengthening.

Procedure

Preparation of a metallic substrate will include the following.

1 **Remove surface coatings (such as paint), scale and existing corrosion products**
   Wire brushing, chipping and wet-blasting can be used to remove surface layers. Care should be taken with cast iron, since it is brittle, and percussive tools (including hammer drills) should not be used on cast iron structures.

2 **Degrease with a suitable solvent**
   An appropriate solvent or detergent should be used sparingly and carefully to ensure that the grease is removed and that widespread contamination of the substrate does not occur. The solvent can be applied by brush and must not be reused. Further detergent solution or alkaline cleaner is often recommended after degreasing to remove dirt and inorganic solids.

3 **Grit-blast**
   Dust and sanding debris should be removed using a vacuum cleaner, although brushing or an oil-free air blast might also be used. Solvent cleaning should not be undertaken after abrasion, as this only partially removes the debris and spreads the remainder over the surface. Water and detergent could be used, provided the substrate is thoroughly rinsed and dried.

4 **Dry the surface**
   Hot air should be used to dry the substrate, as necessary.

5   Chemical etching
The oxide layers that form on aluminium, galvanised steel or stainless steel require acid-etching, which must be followed by neutralisation of the etching products. The chromic oxide layer on stainless steel, in particular, requires the selection of a suitable chemical treatment. Chemical etching is not required with steel or cast iron.

6   Apply primer/adhesion promoter
The primer should be applied by brush or roller to seal the exposed surface of the substrate and promote adhesion. This should be carried out as soon as practicable after cleaning, and always within two hours of cleaning.

7   Produce flat surface profile
The surface of the existing structure must be almost flat and this should be checked using a straight edge. The allowable deviation from a flat surface should be specified by the designer (by considering the resulting peel stress, Section 6.6.4), and should not exceed 5 mm under a 1 m straight edge.

Filler adhesive (or adhesive putty) of adequate strength should be used to fill any irregularities in the surface of the substrate. These irregularities include blowholes in the surface of the cast iron, and the effects of corrosion. For minor surface irregularities, it may be possible to rely upon the adhesive to smooth out any irregularities.

Where the strengthening material wraps around a corner, the radius of the edge should be a minimum of 15 mm to avoid damage to the fibres. This can be achieved by grinding the existing structure (provided that modification of the existing structure is acceptable) or by further adhesive filling.

8   Carry out pull-off tests to check surface preparation
A series of pull-off tests (Section 7.1.1) should be conducted once surface preparation of the substrate (including any priming) has been completed. A minimum of three pull-off tests should be undertaken, but this should be increased where the quality of the surface is highly variable.

7.3.3   Preparation of the strengthening material

In many metallic structures the surface is undulating, and it may curve inwards or outwards. If preformed strengthening is being used, it is advisable to measure the profile of the existing members and manufacture the FRP laminates to match as closely as possible the profile of the beams, while satisfying maximum curvature requirements.

Before use, the FRP strengthening material should be inspected for handling damage, twisting and visible flaws in the resin or fibre lay-up.

If the FRP is not supplied in the required lengths, it will first be necessary to cut the strengthening material to size. All the composite components should be cut to size before installation. Cutting to size of the strengthening products can be carried out using simple tools (Figure 7.5). Sawing is not recommended, as this can cause separation of the fibres within the matrix.

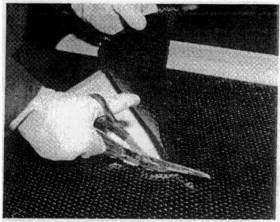

Figure 7.5   *Cutting the FRP strengthening material (courtesy Concrete Repairs Ltd)*

As in the case of the metallic substrate, a clean, chemically active surface layer must be exposed on the surface of the composite material. Composite plates normally include a peel-ply, a sacrificial layer of glass-fibre and polymer material that is placed on the surface during the laminating process to protect the laminate. Immediately before the adhesive is applied to the composite, the peel-ply is removed, leaving a clean surface with the correct amount of surface roughness for a good bond. No further surface preparation is required.

If the composite does not contain a peel-ply, the surface of the composite must be lightly abraded on the side that is to be bonded. Care should be taken not to over-abrade the composite, because this will damage the composite fibres. West (2001) found that the use of abrasive pads provided a suitable surface for bonding. For larger specimens, West recommends bead-blasting, using glass beads of diameters in the range 180–300 microns, a blast pressure of 140–200 kPa, and with the nozzle 50–75 mm from the surface of the CFRP.

The abrasion products should be removed, using a dry cloth, brush, vacuum cleaner or dry air jet. The surface should be wiped with a solvent to remove any residual dust and grease. If the manufacturer specifies a primer coat for the composite, this should be applied immediately before applying the adhesive, or in accordance with the manufacturer's specifications.

Surface preparation of the composite is not required for wet lay-up or *in situ* pre-preg applications.

7.4   APPLICATION OF THE STRENGTHENING MATERIAL

The time available to apply the strengthening material is governed by the pot-life of the adhesive. To ensure smooth installation of a strengthening system, all personnel should be clear about the process to be followed. Before starting work, all tools and temporary works should be prepared and readily available, and the composite should be laid out on a bench or other stable work surface.

During installation, the work area must be kept free of dust and other potential contaminants. This can usually be achieved by thorough cleaning of the workspace before installation, by tenting the work area, and by making sure that contaminants are not allowed to enter at the entrances to the workspace. A small amount of dust is generated if the surface of the strengthening material requires abrasion during the installation process, and this should be collected immediately using a vacuum cleaner.

7.4.1   Preparation of the adhesive

The adhesive will usually be supplied in two parts, which are mixed on-site as required. This should be carried out strictly in accordance with the manufacturer's instructions. The proportions of the two components must be measured accurately, and all containers and tools should be kept clean and in good condition to avoid contamination. If the adhesive is supplied pre-batched, it must not be mixed in smaller quantities than full pots.

Careful note should be taken of the pot-life of the adhesive (see, for example, the list of adhesives in Appendix 3). Any adhesive left after this time should be discarded and a fresh batch mixed. The pot-life of an adhesive depends upon its temperature; the adhesive manufacturer will state pot-lives at a variety of temperatures. The temperature of the adhesive is not just determined by the ambient temperature, because the adhesive generates exothermic heat as it hardens. The manufacturer should provide limits on the amount that can be mixed in one batch.

The adhesive can be mixed mechanically (Figure 7.6). High-speed mixing should be avoided, since this can generate too much heat and can also lead to entrainment of air bubbles that can create defects in the bonded joint.

Figure 7.6   *Mixing the adhesive (courtesy Sika Ltd)*

An alternative to batch mixing of the adhesive is the use of a compressed air gun with a mixing nozzle. This allows the adhesive to be mixed as it is applied. The gun can use either cartridges (which require frequent changing) or can draw from larger volumes of adhesive. Again, the gun should be properly cleaned and maintained, and all adhesive should be applied in the works before the expiry of its pot-life.

7.4.2 Preformed (pultruded) plate or strip strengthening systems

With strip or plate composite strengthening methods (Section 3.2.2), the adhesive should first be applied to both the surface of the existing structure and to the surface of the FRP. The adhesive can be applied to the existing structure by hand trowel or spatula, using plastering techniques.

A greater thickness of adhesive should be applied than required in the finished joint, and care should be taken to minimise air voids. It is important to apply the adhesive to the FRP with a triangular profile, with the maximum thickness at the centre of the plate, tapering down to the edges. This will help prevent air voids being formed. One method of achieving a triangular profile is to use a shaped spreader, as shown in Figure 7.7.

The finished adhesive thickness should be in the range 3–5 mm, unless otherwise specified by the designer. The sensitivity of FRP strengthening to the adhesive thickness is discussed in Section 5.3.1.

Figure 7.7 *Applying the adhesive to a strengthening plate (Concrete Society, 2000)*

The composite plate is now offered up to the existing structure (Figure 7.8). Even pressure is then applied to the FRP using a roller to give a consistent thickness of bonded joint and to expel excess adhesive. Rolling should work from the centre-line of the plate towards its edges, and from mid-span towards the ends, so that any air voids are expelled. The designer should specify acceptable range of minimum and maximum adhesive thicknesses.

Figure 7.8   *Applying an FRP strengthening plate (courtesy Sika Ltd)*

A large amount of adhesive can be squeezed out as part of this process, indicating proper adhesive coverage. However, excessive pressure must not be used, as this would result in a reduced bond line thickness.

Temporary support may be required to hold the strengthening plate in position until the adhesive has cured. Temporary support is required where:

- the self-weight of the FRP plate cannot be supported by the uncured adhesive
- the bonding surface is not flat (for example, where pieces of strengthening material need to cross)
- live loads are applied to the structure while the adhesive cures.

Mechanical clamps and wooden spreaders attached to the flange of a beam can be used as temporary support. (Screw clamps can also be used to adjust the thickness of the bond line). The spacing of supports should be such that there is no sag of the strengthening material between supports, and will depend on the specific strengthening scheme (and in particular, the thickness of strengthening plate). Typically, supports will be needed at 300 mm centres, or at such centres as are compatible with the thickness of the plate. Clamps also apply some pressure to the adhesive, ensuring a full-strength joint.

Excess adhesive should be removed using scrapers, cloths and careful use of solvents. Around the ends of plates, a spew fillet (flatter than 45° to the horizontal) should be left along the edges of the plates to alleviate the high local stresses and help seal against moisture and chemical ingress along the bond line.

7.4.3   Wet lay-up (hand lamination) strengthening systems

Where a wet lay-up strengthening system (described in Section 3.2.1) is applied by hand, the adhesive should first be evenly applied to the existing structure using a roller or brush. Dry fibre can either be impregnated with resin (using an impregnation machine), or applied in its dry state (partially pre-impregnated fibres are also available). The fibres are then rolled into the adhesive, to bring the fabric into intimate contact with the adhesive, expel trapped air and remove any wrinkles in the FRP material. Additional layers of adhesive and fabric can be applied by roller, finishing with a top coat of adhesive to encapsulate and protect the strengthening material.

Particular attention should be given to the installation of wet lay-up FRP in highly stressed, critical regions of the strengthening scheme, to ensure that the FRP is properly impregnated and consolidated.

Large quantities of wet resin are used during the wet lay-up process, and the operatives need to work close to this resin for extended periods of time. It is vital to protect operatives from the volatile emissions from the resin. The workspace must be well ventilated, and the operatives equipped with appropriate personal protective equipment (see Section 7.2.1). Operatives should also wear suitable protective clothing to protect against dermatitis caused by skin contact with the resin.

7.4.4 Vacuum bag moulding and vacuum infusion methods

Vacuum bag moulding can be used for further consolidation of strengthening materials that have been applied by the wet lay-up method and to remove voids.

Vacuum infusion allows dry fibres and mats to be applied directly to the existing structure, with the vacuum used to draw in the resin.

Both techniques are described in Section 3.2.1.

7.4.5 Tests to assess the quality of the installation

Visual and sonic inspection

The installed strengthening scheme should be visually inspected immediately after assembly, and again after the adhesive has hardened and any temporary supports have been removed. Section 8.1 (and in particular Table 8.1) gives further details on post-installation inspection.

It is possible to repair local defects in non-critical regions (Section 8.2), but widespread defects will require the strengthening material to be removed and the whole strengthening process, including surface preparation, to be repeated.

Test specimens

Tests should be conducted on specimens prepared throughout the installation process (Section 7.1.1). It is recommended that three of each type of test specimen be prepared every morning and afternoon that work is carried out. These specimens should be tested at an independent testing centre.

If preformed strengthening material is being used, the material supplier will already have carried out tests to verify that the strengthening material has the required characteristics. If the composite is formed on site (such as by wet lay-up or vacuum infusion methods), test samples must be produced during installation. In all cases, adhesive test samples should be prepared.

It is beneficial to identify installation problems early in strengthening works, before the principal works are installed. Hence it is prudent to prepare lap-shear test specimens (Section 7.1.1) while the strengthening work is being carried out and to test these as soon as possible after the adhesive has cured. If the strength of the joint indicated by these tests is inadequate, the strengthening material must be removed and the installation process repeated.

The lap-shear specimens reflect the quality of the installation process, including:

- correct mixing of the adhesive
- adequate surface preparation of the composite
- suitable environmental conditions at the time of installation
- correct application of the adhesive and strengthening.

Tensile strength, flexural modulus and glass transition temperature specimens should also be measured for the adhesive while on site, according to the standards listed in Table 7.1.

Non-destructive testing

If there is reason to doubt the integrity of the adhesive bond, non-destructive testing techniques should be used, as discussed in Section 8.4.

7.5   CURING

The temperature of the structure (as well as the ambient air temperature) is critical for a high-strength bond to be achieved. The first 24 hours of curing are the most important. As Figure 3.1 shows, however, although the adhesive hardens rapidly, it needs to continue to cure for several days to achieve its full mechanical properties, particularly when the adhesive cures at ambient temperature. It will usually be necessary to leave any temporary supports in place for the whole of this period.

The rate of cure is temperature-dependent for any adhesive, and the manufacturer will specify a temperature window for cure to ensure a full-strength joint. Hot-cured adhesives (for example, in partially cured pre-impregnated strengthening materials) require an elevated temperature during cure, and the manufacturer will again specify the required temperature window. The air temperature around the structure can be controlled using heaters (as discussed in Section 7.2.3). Alternatively, electric heating blankets can be applied around the strengthening.

The metallic elements conduct heat to the external environment and can be much cooler than the air temperature in the enclosure. For example, heating a cast iron bridge before strengthening raised the air temperature within the tent to 35°C, but the temperature of the cast iron remained at half this value. If the temperature of the metal is expected to be critical for curing the adhesive because of the low external ambient temperature, the temperature of the metal beams should be monitored at positions along their lengths using thermocouples inserted in the adhesive layers adjacent to the metallic substrate.

Excess humidity can be detrimental to the durability of a bonded joint, and the humidity should also be controlled while the adhesive cures. The bonded joint should therefore be protected from contact with water during the curing period.

Dynamic live loads applied during the curing process can prevent a full-strength adhesive joint forming. The structure should not be loaded until the adhesive has cured, unless the design process has taken this into account (Section 6.4.2).

7.6   PERMANENT LOAD TRANSFER METHODS

7.6.1   Prestressing

Prestressing the composite adds an additional stage to the installation process, which must be carried out within the working life of the adhesive. Any system for prestress application must therefore be quick and easy to handle on site.

The prestressing system will usually comprise:

- prestressing jacks
- reaction arrangements for the prestress load
- gripping tabs (through which the prestress is applied to the FRP plates)
- permanent anchorages.

The end anchorages are particularly important details in the system. Gripping tabs are typically bonded to the FRP strengthening material and transfer the jack load into the FRP. The gripping tabs and jacking arrangement must ensure that the strengthening material is correctly aligned with the existing structure. This requires a very stiff stressing system, particularly since one side of the FRP is almost in contact with the existing structure.

Section 4.2.1 describes the importance of permanent anchorages to carry the high stresses at the ends of a strengthening plate. These must be tailored to suit the form of the structure being strengthened.

It is important to carry out the stressing within a suitable ambient temperature window, in order to avoid damage of the strengthening material due to stress concentrations. Bonded tabs are susceptible to embrittlement at low temperatures, which can lead to failure. It is recommended that a bonded end anchorage be supplemented by a mechanical connection.

7.6.2   Load-relief jacking

Load-relief jacking (described in Section 4.2.2) enables a proportion of the permanent stresses to be transferred from the existing structure into the strengthening material.

If the jacks bear directly on the surface to be strengthened, they will restrict the area over which the FRP strengthening can be applied, since the jacks must be in place throughout the strengthening operation. If this limitation creates a problem, alternative methods of lifting the beams to be strengthened must be sought in which the beam lifting points are placed away from the surfaces to be strengthened.

It is important to check that the metallic structure can accommodate the jacking forces without damage. For example, without temporary bracing, the web of an I-section girder may have insufficient bearing capacity at midspan to accommodate a significant jacking force applied there.

The load transfer operation requires careful monitoring to avoid over-stressing the existing structure. The applied load and resulting deflections should be recorded, together with strain-gauge measurements in the existing structure.

7.7 FINISHES

Where the metallic substrate is left around the strengthening material, it will require protection against corrosion. The paint should be extended over the exposed edge of the adhesive joint, as this adds a further layer of protection, sealing the adhesive joint against moisture and chemical ingress. The paint system should be chemically compatible with both the adhesive and the strengthening system used. The requirements of an epoxy-based paint system and a structural epoxy adhesive are different, therefore using the adhesive as a corrosion protection layer is not recommended.

Wherever possible, CFRP strengthening plates should not be painted, as this inhibits their inspection. However, depending on their exposure conditions, AFRP strengthening plates may need to be painted, to protect them from unacceptable ultraviolet degradation.

The FRP strengthening material is at risk of damage during the installation of finishes and fittings, or later modifications to the structure. For example, holes must not be drilled into the composite to attach light fittings. Leaving the strengthening material exposed helps to highlight the presence of a different material. However, it is advisable to fix clearly visible identification and warning plates to the strengthened area. TR57 (Concrete Society, 2003) shows example warning labels.

7.8 BIBLIOGRAPHY

Concrete Society (2000). *Design guidance for strengthening concrete structures using fibre composite materials.* Technical Report 55, Concrete Society, Crowthorne

Concrete Society (2003). *Strengthening concrete structures with fibre composite materials: acceptance, inspection and monitoring.* Technical Report 57, Concrete Society, Crowthorne

The Construction (Design and Management) Regulations 1994. SI 1994/3140, HMSO, London

Control of Substances Hazardous to Health Regulations 1999. SI 1999/437, Stationery Office, London

fib (2001). *Externally bonded FRP reinforcement for RC structures.* fib bulletin 14, Fédération internationale du béton, Lausanne

Health and Safety at Work etc Act 1974. 1974 c. 37, HMSO, London

IStructE (1999). *A guide to the structural use of adhesives.* SETO, London

Mays, G C and Hutchinson, A R (1992). *Adhesives in civil engineering.* Cambridge University Press

Moy, S S J, ed (2001). *FRP composites – life extension and strengthening of metallic structures.* ICE design and practice guide, Thomas Telford, London

West, T D (2001). *Enhancement to the bond between advanced composite materials and steel for bridge rehabilitation.* CCM Report 2001-04, University of Delaware Center for Composite Materials

British Standards

BS 1881-207:1992. *Testing concrete. Recommendations for the assessment of concrete strength by near-to-surface tests*

BS 2782-3:Methods 320A to 320F:1976. *Methods of testing plastics. Mechanical properties. Tensile strength, elongation and elastic modulus* [obsolescent]

BS 5350-C5:2002. *Methods of test for adhesives. Determination of bond strength in longitudinal shear for rigid adherends*

BS 6319-3:1990. *Testing of resin and polymer/cement compositions for use in construction. Methods for measurement of modulus of elasticity in flexure and flexural strength*

BS 6319-8:1984. *Testing of resin and polymer/cement compositions for use in construction. Method for the assessment of resistance to liquids*

BS 7079-0:1990. *Preparation of steel substrates before application of paints and related products. Introduction*

BS EN 1770:1998. *Products and systems for the protection and repair of concrete structures. Test methods. Determination of the coefficient of thermal expansion*

BS EN 14022:2003. *Structural adhesives. Determination of the pot life (working life) of multicomponent adhesives*

BS EN ISO 527-4:1997/BS 2782-3:Method 326F:1997. *Plastics. Determination of tensile properties. Test conditions for isotropic and orthotropic fibre-reinforced plastic composites*

BS EN ISO 527-5:1997/BS 2782-3:Method 326G:1997. *Plastics. Determination of tensile properties. Test conditions for unidirectional fibre-reinforced plastic composites*

BS EN ISO 899-1:2003. *Plastics. Determination of creep behaviour. Tensile creep*

BS EN ISO 899-2:2003. *Plastics. Determination of creep behaviour. Flexural creep by three-point loading*

BS EN ISO 4624:2003/BS 3900-E10:2003. *Paints and varnishes. Pull-off test*

BS EN ISO 14125:1998. *Fibre-reinforced plastic composites. Determination of flexural properties*

BS EN ISO 14126:1999. *Fibre-reinforced plastic composites. Determination of compressive properties in the in-plane direction*

BS EN ISO 14129:1998. *Fibre-reinforced plastic composites. Determination of the in-plane shear stress/shear strain response, including the in-plane shear modulus and strength by the ± 45° tension test method*

BS EN ISO 14130:1998. *Fibre-reinforced plastic composites. Determination of apparent interlaminar shear strength by short-beam method*

BS ISO 1268-1:2001 to 1268-9:2003. *Fibre-reinforced plastics. Methods of producing test plates* [various subtitles]

BS ISO 11357-2:1999. *Plastics. Differential scanning calorimetry (DSC). Determination of glass transition temperature*

BS ISO 11359-2:1999. *Plastics. Thermomechanical analysis (TMA). Determination of coefficient of linear thermal expansion and glass transition temperature*

BS ISO 15310:1999. *Reinforced plastics. Determination of the in-plane shear modulus by the plate twist method*

8 Inspection and maintenance

A strengthened structure requires inspection (a) immediately after the strengthening operation and (b) at regular intervals during the structure's life. This chapter outlines a typical inspection and maintenance regime for an FRP-strengthened metallic structure.

Section 8.1 identifies possible defects that should be checked for, and can be used as a checklist during an inspection. NDT methods are briefly outlined, together with performance monitoring systems. Anticipated maintenance requirements are covered in Section 8.2, whilst various options for repairing FRP strengthening systems are described in Section 8.3.

The Concrete Society publication TR57 (Concrete Society, 2003) gives a detailed description of inspection and maintenance requirements for externally bonded FRP applied to concrete structures, which can be used as a further source of reference for metallic structures strengthened using FRP.

8.1 WHAT TO LOOK FOR DURING INSPECTIONS

The existing structure and the strengthening system should be inspected at suitable regular intervals. Regular inspections are particularly important for FRP strengthening schemes, since this is a relatively new technology for which there is a lack of long-term experience.

Initially, the strengthened structure should be *closely* inspected every six months, to check that the strengthening system is functioning satisfactorily. In the first six years, inspections should be carried out every six months. Thereafter, yearly visual inspections, and close inspections every six years, will normally suffice.

General advice on the inspection and appraisal of structures (including health and safety considerations) is given in the Institution of Structural Engineers publication *Appraisal of existing structures* (IStructE, 1996). Specific advice is produced by the major bridge owners. BD 21/01 (Highways Agency, 2001a) is one such document applicable to highway bridges, and is publicly available.

The structure's health and safety file must include a description of the strengthening scheme and define the inspection requirements (see Section 7.1.2). These records are essential when appraising a structure: for example, the lay-up of fibres within the FRP can otherwise only be determined by burning off the resin. The results of all inspections should be added to the health and safety file.

Tables 8.1 and 8.2 list possible defects and inspection techniques, which should be considered during an inspection. Table 8.1 lists defects that should be checked for during the end of installation inspection and Table 8.2 relates to in-service inspections.

Table 8.1   *Possible defects and inspection techniques for inspection at the end of installation*

Component of the strengthened structure	Possible defect	Inspection technique
Position of strengthening	• Incorrect placement of FRP • Misalignment of the strengthening	• Visual inspection
Surface of FRP	• Manufacturing defects, including surface blisters) • Dry spots where resin has not saturated fibres • Exposed fibres (especially at cut edges)	• Close visual inspection.
Within the FRP	• Internal delamination	• Tapping to locate hollow regions • NDT techniques
The edge of the adhesive layer	• Debonding of the FRP • Integrity of the spew fillet of adhesive around the edge of the strengthening material	• Visual inspection
The adhesive layer beneath the FRP	• Incorrect adhesive thickness • Delamination • Air voids • Grease and other foreign matter in the adhesive	• Visually inspect and measure uniformity of adhesive thickness • Tapping to locate hollow regions • NDT techniques • Contamination of the adhesive is particularly difficult to detect in the finished joint, therefore it should be controlled during the application process
Monitoring system (if installed)	• Change in structural response of the strengthening system (for example during the transfer of prestress)	• Check for unexpected changes in the output of the monitoring system (which might include embedded strain gauges or optical surveying targets)

The size of defect that is allowable within a strengthened structure depends upon the method of strengthening, and the position of the defect within the strengthened member. For example, in the highly stressed adhesive near the ends of a strengthening plate only a small defect will be acceptable, whereas towards the middle of the strengthening plate, where the shear stress is low, a larger-sized defect may be acceptable. Particular attention should thus be given to critical regions of the structure during the inspection (such as the ends of the plates and any anchorage arrangements).

A defect should thus be assessed according to the stress level at its location within the structure, and its proximity to other defects. The size of suggested allowable defects at various positions within the structure should be specified during design, and included in the structure's health and safety manual. For example, Figure 8.1 shows a decision tree that could be used to assess the significance of defects in the adhesive bond.

Table 8.2 *Possible defects and inspection techniques for in-service inspections*

Component of the strengthened structure	Possible defect	Inspection technique
Original structure	• The original structure must be inspected for further degradation. Damage to the original structure can lead to overstressing of the strengthening material	• Refer to IStructE (1996) and BD 21/01
Surface of FRP	• Loss of surface resin. • Crazing, pitting, scoring or blistering of resin • Surface abrasion • Exposed fibres	• Close visual inspection
Within the FRP	• Internal delamination	• Tapping to locate hollow regions • NDT techniques
The edge of the adhesive layer	• Debonding of the plate • Crazing and other degradation of the resin • Integrity of the spew fillet of adhesive around the edge of the strengthening material	• Visual inspection
The adhesive layer beneath the FRP	• Delamination • Air voids • Grease and other foreign matter in the adhesive	• Tapping to locate hollow regions • NDT techniques • Contamination of the adhesive is particularly difficult to detect in the finished joint, so it should be controlled during the application

CFRP strengthening systems	• Galvanic corrosion	• Galvanic corrosion may be indicated by high levels of local corrosion of the metal. Check for anything bridging between the CFRP and metal substrate, eg:
		  − bolts and anchorages that have not been insulated
		  − corrosion products close to the CFRP

Protective finishes	• Damage to coatings:
	  − paints or resin layers
	  − fire proofing
		• Visual inspection

Monitoring system (if installed)	• Change in structural response of the strengthening system	• Check for unexpected changes in the output of the monitoring system (which might include embedded strain gauges or optical surveying targets)
Test specimens (prepared at the same time as the strengthening operation)	• Degradation of strengthening materials	• If specimens were prepared for future testing as part of the installation, these should be tested at the specified interval, or if other aspects of the inspection give cause for concern. Degradation of the test specimens may indicate degradation of the materials used to strengthen the structure
Associated works	• Warning or identification labels absent	• Visual inspection
General	• Fire damage, impact damage, vandalism	• Visual inspection
General	• Damage due to other work carried out in the region of the strengthening material	• Visual inspection (eg check that holes have not been drilled through the FRP strengthening material)
General	• Previous repairs to the strengthening system	• Particular attention should be given to regions where the strengthening system has been repaired

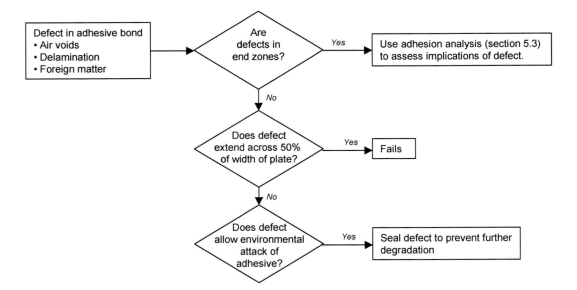

Figure 8.1 *Suggested decision tree for determining the significance of a defect in the adhesive*

Table 8.3 gives an indication of possible allowable defect sizes for wet lay-up strengthening. However, the defects allowed in any particular project should be assessed as described above. It should be noted that whilst it is useful to know the allowable size of defect to aim for, it is not always possible to inspect for the required defect size.

8.2 REPAIR STRATEGIES

Repair strategies should be considered when the strengthening scheme is designed, and the details recorded in the structure's health and safety file.

Repair can be affected by replacing damaged regions of a strengthening system with new material. The materials used in the repair must be compatible with the original material (both chemically and mechanically). This will usually require the use of a composite with the same fibre orientation, fibres, resin and/or adhesive.

Any damage that leaves the fibres or the bonded surface exposed to moisture or chemical attack must be addressed at the earliest opportunity to avoid further degradation of the strengthening system.

Surface resin damage

Damage to the surface of FRP strengthening materials can be repaired by applying a coat of resin, after carefully cutting away any ineffective resin, abrasion, and solvent cleaning (as in the installation chapter). If any fibres are damaged, the reduction in strength of the composite should be assessed, to establish whether composite patching is required to ensure continuity of the load path.

Table 8.3   *Suggested allowable defect sizes in wet lay-up FRP strengthening*

Defect	Suggested allowable limits
FRP – resin-rich surface layer	
Cracks	No appreciable incidence permitted
Foreign matter	Maximum 10 mm long, 2 mm deep
Pits/scores	Maximum 10 mm diameter, 2 mm deep
Surface geometry	No step changes in thickness greater than 2 mm
FRP – body	
Blisters	Maximum 5 mm diameter, 2 mm high
Delamination	No appreciable incidence permitted
Cracks	No appreciable incidence permitted
Exposed cut edges/fibres	No appreciable incidence permitted
Foreign matter	No appreciable incidence permitted
Pits/scores	Maximum 5 mm diameter, 2 mm high
Porosity	No appreciable incidence permitted
Surface geometry	No step changes in thickness greater than 2 mm
Unimpregnated/dry fibre	No appreciable incidence permitted
FRP – substrate interface	
Delamination	No appreciable incidence permitted
Dry spots	Max 10 per m^2; total no greater than 100 mm^2 in area
Foreign matter	No appreciable incidence permitted
Surface geometry	No step changes in thickness greater than 2 mm
Surface corrosion	No appreciable incidence permitted

Debonding

A low-viscosity resin can be injected between the FRP and the metallic structure to repair an adhesive joint. Advice should be obtained from the material supplier on a suitable resin, which must be chemically and mechanically compatible with the existing strengthening components. Care must be taken to avoid excessive injection pressure, which can cause further delamination damage.

The engineer must be satisfied that the bonding surfaces are not contaminated, for example, by corrosion of the substrate. A suitable resin should be drawn into the debonded area under vacuum. Resin injection should be avoided, as it can lead to further debonding around the existing defect. The effectiveness of the repair should be assessed by tapping the surface of the composite to check that the strengthening material does not sound hollow.

Resin injection should be followed by replacement of the adhesive fillet around the edge of the bond.

Composite patching

Where the composite fibres have become damaged (for example, due to fire or impact) it may be possible to apply a composite patch.

Before applying the patch, an analysis should be undertaken to determine its effectiveness. The damaged area should be treated as a hole in the strengthening material and the resulting stress concentration established. The patch provides an alternative load path and so reduces the stress concentration. It is unable to carry any of the permanent loads carried by the structure when the patch is applied, however.

The damaged material is cut out from the surrounding strengthening material. New material is then added, either in the form of a prefabricated laminated plate or by a wet lay-up process (possibly accompanied by a vacuum bag consolidation process). Where possible, additional layers of composite can be built up above the surface of the existing strengthening material.

Replacement

In extreme circumstances, it may be necessary to replace a piece of strengthening material completely. Removal of the existing material is relatively straightforward and can be effected by driving wedges along the adhesive layer – if necessary, in conjunction with prior heating of the adhesive to above its T_g value. Care should be taken when using wedges on a brittle cast-iron structure to ensure that excessive tensile stress in the cast iron does not cause failure. The strengthening system can then be reapplied by following the original installation procedure.

If the composite carries some permanent load (originally introduced by prestressing or load-relief jacking), special care must be taken during its removal. If necessary, the permanent load can be relieved from the composite by reversing the installation process, by detensioning in a controlled manner.

8.3   ANTICIPATED MAINTENANCE REQUIREMENTS

The anticipated maintenance requirements for an FRP strengthened structure are minimal. Where the layer of surface resin is lost (due to abrasion or environmental degradation), this can easily be replaced by recoating using a compatible resin (as described in the previous section).

The existing metallic structure will require continuing maintenance, in the same manner as before it was strengthened. During this work, care should be taken not to damage the FRP strengthening materials. For example, the paint on a structure should be renewed before it fails, so that it is not necessary to grit-blast the structure, which would require protection of the strengthening material.

8.4　NDT TECHNIQUES

Various non-destructive testing (NDT) techniques exist for examining composite strengthening materials and bonded joints, although they are not generally readily used on site. There is a lack of practical, well-developed NDT techniques for inspecting FRP-strengthened metallic structures, but they are likely to become more useful in the future.

Different NDT methods are capable of detecting different types of defect (as described below). The type of NDT adopted should be chosen to suit the nature of the defect that is of concern.

The limitations of the NDT technique chosen must be recognised. In particular, while it is possible to detect air voids in the adhesive, no NDT technique exists to detect poor adhesion due to surface dust or grease in the adhesive layer.

Whatever technique is adopted, it must be capable of detecting the size of defect that is of interest (for example, the imperfection sizes shown in the tables above). It is important to recognise that the results of NDT may be confused by flaws in the underlying substrate.

Some of the available NDT methods are listed below.

Acoustic inspection

Ultrasonic techniques are most commonly used to inspect composite materials. Defects (such as delamination, debonding and voiding) interrupt the path of the ultrasound wave and affect its transit time. Specialist equipment and personnel are required.

The *acoustic pulse velocity* technique employs lower-frequency pulses and has a lower resolution, but works in the same manner as ultrasonic inspection.

Acoustic emission

Acoustic emission is a live global inspection technique used to monitor the composite structure while under load. It is not normally capable of distinguishing specific failure modes or events, but can be used to assess the overall damage state of the composite, assuming a suitable reference point has been defined. If a sufficient number of detectors is employed, then it should be possible to detect and locate specific failure events. Acoustic emission is useful for ongoing monitoring of the performance of a structure.

Radiographic techniques

Radiographic techniques detect voids, inclusions and other defects within a material using X-ray or gamma-ray transmission.

Infrared thermography

Infrared imaging techniques can be used to identify small temperature differences on the surface of a structure as it heats up or cools down (due to natural daily or seasonal variation). These temperature differences indicate interruptions to the heat flow, such as voids within the composite or adhesive.

Infrared thermography is a relatively new technique. It has been used on other structures, but experience with FRP strengthening systems is still very limited. Passive infrared thermography analyses the temperature patterns of the structure due to ambient exposure while active thermography does the same for exposure to controlled external heat sources.

Laser shearography

Laser shearography is a rapid inspection technique that is used for inspecting large areas such as composite structures (lifeboat hulls) and pressure vessels in real time. Shearography senses out-of-plane surface displacement in response to an applied load, the surface displacements being indications of flaws/defects within the material. It is capable of detecting delaminations within the composite material or at the interface between the composite and the substrate. The most effective means of applying the load is using surface vacuum techniques. Other methods of load application include surface heating and, where practicable, internal pressure.

8.4.1 Performance monitoring

Instrumentation can be incorporated as part of the installation process to monitor the behaviour of the strengthened structure. It can be applied both to the existing structure and to the strengthening materials. This can be useful both during the installation process (particularly during load-relief jacking, or prestressing operations), to check for changes in the performance of the structure in the longer term, and hence monitor the durability of the applied strengthening system.

Amongst the instrumentation that can be installed on a structure are:

- foil strain gauges
- fibre-optic strain gauges, which can be used to measure strain at several points along their length and can be embedded within the structure
- deflection gauges
- load cells.

The acoustic emission technique (described above) can also be used to monitor the performance of a structure.

8.5 BIBLIOGRAPHY

Concrete Society (2003). *Strengthening concrete structures with fibre composite materials: acceptance, inspection and monitoring.* Technical Report 57, Concrete Society, Crowthorne

Highways Agency (2001a). *The assessment of highway bridges and structures.* BD 21/01 (DMRB vol 3, sec 4, pt 3), Stationery Office, London

IStructE (1996). *Appraisal of existing structures.* 2nd edn, SETO, London

Moy, S S J, ed (2001b). *FRP composites – life extension and strengthening of metallic structures.* ICE design and practice guide, Thomas Telford, London

9 Owners' considerations

This chapter describes management issues that will generally be of interest to the client or owner of a metallic structure that is in need of strengthening. Section 9.1 outlines the procurement process. An economic assessment of externally bonded FRP strengthening systems with respect to alternative strengthening methods should be carried out early in the procurement process and this is described in Section 9.2. Many metallic structures are of historic significance and Section 9.3 covers the use of FRP materials to strengthen heritage structures. Awareness of the environmental implications of structural work is increasing and the environmental impact of an FRP strengthening scheme is discussed in Section 9.4.

9.1 PROCUREMENT

External FRP strengthening of metallic structures is a relatively young technology, hence a traditional route is preferred for procuring strengthening works, as shown in Figure 9.1. However, it is strongly recommended that both the contractor and materials supplier are involved early in the procurement process, since they have the best understanding of the materials involved, how they work and how they are to be installed.

The need for a structure to be strengthened will be determined during a structural assessment, which might be prompted by the inspection regime or a proposed change in use of a structure. If strengthening is required, a consulting engineer will be appointed to develop conceptual designs for alternative strengthening methods, including external FRP strengthening. (The consultant might previously have been involved during the structural assessment stage.)

During the conceptual design stage, the FRP option is compared with alternative strengthening methods. Section 9.2 (economic considerations), Section 9.3 (for heritage structures), and section 9.4 (environmental considerations) provide guidance on assessing the appropriate strengthening technique.

If externally bonded FRP strengthening is chosen, the consultant should continue with the detailed design. The designer of the FRP strengthening scheme must have previous relevant design experience. It will usually be necessary to employ a specialist consultant for this purpose.

Tender documents can be prepared on the basis of the detailed design to allow a contractor to be appointed. These should include a detailed specification of the materials and workmanship requirements.

The strengthening material is usually procured through a specialist contractor who is responsible for the installation. The contractor must have previous, relevant experience of applying FRP strengthening systems, preferably to metallic structures. The competence of the contractor should be demonstrated by providing evidence of operative training, or by documentary evidence relating to previously completed relevant projects. The contractor should operate a quality assurance system, accredited and audited in accordance with BS EN ISO 9002:1994. The contractor must submit a proposed method statement to the client, which must be approved before any strengthening works begin.

It is essential that the installation work should be supervised by suitably trained staff who have previously successfully supervised and managed the FRP strengthening of metallic structures. The installation operatives should also have received appropriate training from the supplier of the strengthening system and have passed qualification tests. Personnel who have previously applied FRP strengthening material to concrete structures should be aware of the differences between the two substrate materials.

It is strongly recommended that an industry-wide certification scheme be established to ensure that the operatives and supervisors installing an FRP strengthening scheme have the appropriate training and experience.

Externally bonded FRP strengthening materials are generally sold as strengthening *systems*, rather than individual components. The supplier will usually be able to provide equipment for any handling or processing of the composite that is required, prestressing equipment, and tools for infusion and impregnation (as appropriate). Depending upon the installation process, the supplier will be able to provide either trained personnel or training for the contractor's own operatives. If the materials are not certified by a recognised certification scheme, the supplier needs to provide test data in support of its products.

Before any installation work starts, the contractor must submit a final method statement for the strengthening works, which must be approved by the client. It is strongly recommended that a representative of the designer be engaged to provide supervision during installation, since lack of appropriate supervision can result in costly errors.

There may be significant benefits to be gained from integrating the supply chain and early specialist involvement; this would involve early collaboration between designer, materials supplier, specialist contractor and client.

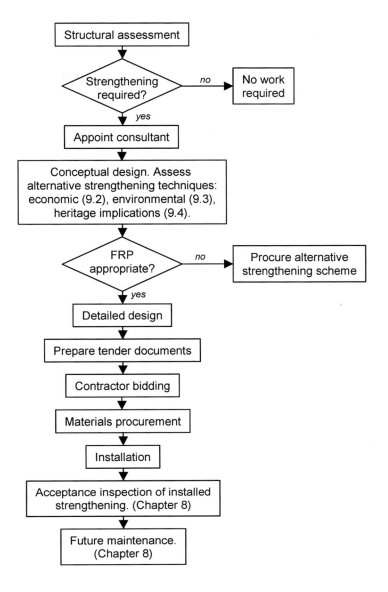

Figure 9.1   *Suggested procurement route for external FRP strengthening*

9.2   ECONOMICS

In the conceptual design phase an externally bonded FRP strengthening solution for a metallic structure should be assessed in comparison with other possible strengthening options available. Some of the strengthening options are listed below.

Alternatives to strengthening

- replacement
- modification to the structural form (for example, by inserting props at mid-span or adding new beams)
- live load reduction (by imposing lane or track closures, weight restrictions or speed limits)
- dead load reduction (for example, by replacing a heavy concrete floor or deck with one made from FRP composite materials).

Strengthening using conventional materials

- external post-tensioning using cables
- fatigue life extension by drilling crack-stopper holes or by local re-welding.
- over-slabbing
- addition of metal to increase the sectional depth (by bolting or welding on extra material)
- local strengthening (for example, the addition of stiffeners to increase local buckling capacity)
- filling or encapsulation in concrete.

Strengthening using FRP materials

- bonded FRP, without prestress
- prestressed FRP plate bonding
- FRP post-tensioning cables.

Non-metallic cables are available that can be used for post-tensioning. These include unimpregnated dry parallel fibre strands and fibres combined with a matrix to produce FRP. Although FRP post-tensioning cables fall outside the scope of this report, they are essentially a variant of conventional post-tensioning, but are significantly more durable than steel tendons (Hollaway and Head, 2001).

The cost of the materials required for externally bonded FRP strengthening systems is comparatively high. The high material cost is generally offset by reductions in the costs of access, temporary works, installation and operation. When comparing the cost of different strengthening options, the following items should be considered:

- the supply cost
- the installed cost
- the annualised life cycle cost.

Bonding a strengthening element to the existing structure is a relatively low-cost method of installation. Alternatives, including steel-plate bonding (Department of Transport, 1994), require holes to be drilled in the existing structure, which leads to additional expense and short-term weakening of the structure.

The installation of externally bonded FRP can often be carried out more quickly and with far less disruption than alternative methods of strengthening. Substantial costs may be associated with disruption, such as those derived from track possession on railways.

FRP strengthening materials are lightweight (especially when compared with steel plate strengthening), and the costs of both access arrangements and lifting equipment will be reduced. Further savings include the shorter time required to erect scaffolding and lower costs of transporting equipment to site.

The need and extent of strengthening work will usually be determined by a structural assessment. An economic FRP strengthening design requires consideration of the whole structure (and should not be based on a required bending moment envelope). It is therefore advantageous to combine the assessment and strengthening phases of the design into a single project. The costs of decommissioning (further discussed below) should be considered in the whole-life cost of a strengthening scheme.

9.3   HERITAGE STRUCTURES

If the structure is of historic significance, interested bodies such as English Heritage, Historic Scotland, CADW or the Department of the Environment in Northern Ireland must be consulted before any work is undertaken.

The design of strengthening schemes for such structures must consider the following requirements.

Reversibility

The strengthening component should be designed for easy removal, allowing the structure to be reinstated to its initial (unstrengthened) state should the strengthening prove to be unsatisfactory.

Minimal structural and visual intrusion

The strengthening system should be non-intrusive structurally and visually. The form of the structure should not be affected and original construction details should not be obscured.

Maximum retention of the existing structure

Wherever possible, the existing structure should be left in place. This includes details such as rivets and bolts.

The appropriate use of FRP strengthening materials, applied such that no damage is caused to the existing structure, can be a beneficial way to extend the life of a heritage structure. Strengthening material is added to the existing structure (leaving the existing members in place), and can be designed to minimise intrusion. In the event of problems, the FRP can be peeled away from the existing structure, providing a reversible solution.

While the strengthening material itself does not usually have a large visual impact on the structure, careful consideration should be given to coatings required for fire protection, and, in the case of a prestressed strengthening system, the anchorage arrangement.

9.4   ENVIRONMENTAL CONSIDERATIONS

Environmental legislation in the UK and Europe is becoming increasingly demanding, with larger taxes and fines resulting from environmentally detrimental activity. Lainchbury and Edwards (2001) describe a framework under which the environmental impact of different construction techniques can be assessed.

It has been estimated that the construction industry produces 34 per cent of the UK's waste. Strengthening a structure allows its life to be extended, instead of demolished and replaced, which makes a worthwhile contribution towards reducing waste and preserving the environment.

The environmental impact of a strengthening scheme can be minimised through design by:

- minimising the volume of products used
- designing the strengthening scheme to be easy to dismantle
- using a reusable or recyclable strengthening material (Collins and Conroy, 2001).

These issues are addressed in subsequent sections.

9.4.1 During installation

Protection of the surrounding environment during the installation process was described in Section 7.2.2. Similar environmental protection measures should be implemented during any repair work.

9.4.2 During service

In normal service, no environmental issues are expected. However, fire, high impact or other destructive actions may release fumes and fine dust particles into the environment. The environmental impact of such events is limited.

Maintenance of the strengthened structure may require removal of some strengthening materials and for additional composite materials and adhesives to be used. The measures described in Section 7.2.2 should be taken to protect the environment.

9.4.3 Removal of the strengthening system from a structure

FRP strengthening materials can be separated from the structure along the line of the adhesive bond using a combination of heating and wedges. Where the composite needs to be cut, the workforce and the environment should be protected from dust and fibrous particles.

Disposal of composites

Once the strengthening material has been removed from the structure, the composite and adhesive must be disposed of in accordance with local regulations.

Composites are not normally considered chemically hazardous to the environment. They do not contain any substance that could leach out to contaminate the groundwater or air. Composites are generally sent to landfill sites, but UK and European legislation encourages the minimisation of waste disposed of in this manner.

The following alternative waste reduction methods are available for polymeric composites, listed in decreasing desirability (Collins and Conroy, 2001).

Waste minimisation

The amount of waste can be minimised by reducing the volume of material applied to a structure. If the strengthening component becomes damaged, it can be repaired locally (by patching), rather than by replacing the composite all along a member.

Reuse

It may be possible to reuse the composite strengthening material in its present form. Uses should be considered outside the structures industry, since it is unlikely that the composite could be reused as structural strengthening.

Recycling

Methods of recycling composite materials are being developed, and pilot FRP recycling plants are active in France, Germany, Italy and the Netherlands (Collins and Conroy, 2001). At the moment, thermoset composites are ground and used as filler material in new composites, asphalt and other specialist polymer compounds.

Fibre recovery by thermal decomposition or chemical degradation is being investigated. Both methods of recycling have environmental implications of their own, and the recovered fibres are unlikely to retain their original strength, so will not be suitable for structural applications.

Incineration with energy recovery

Composites have a high calorific value, which can be retrieved by incineration. Combustion of the matrix, however, produces toxic fumes, and incineration of carbon materials may release fine, electrically conductive particles into the air.

Incineration without energy recovery; landfill

Incineration without energy recovery and landfill should only be considered if none of the previous options are practicable.

9.5 BIBLIOGRAPHY

Collins, R and Conroy, A (2001). "Composites: design for deconstruction, reuse and recycling". In: *Proc NGCC 1st ann conf & AGM, Composites in construction; through life performance, 30–31 Oct, Watford*

Department of Transport (1994). *Strengthening of concrete highway structures using externally bonded plates*. BA 30/94 (DMRB vol 3, sec 3, pt 1), HMSO, London

Hollaway, L C and Head, P R (2001). *Advanced polymer composites and polymers in the civil infrastructure*. Elsevier Science

Lainchbury, J and Edwards, S (2001). "The life cycle environmental impacts of construction products". In: *Proc NGCC 1st ann conf & AGM, Composites in construction; through life performance, 30–31 Oct, Watford*

British Standard

BS EN ISO 9002:1994. *Quality systems. Model for quality assurance in production, installation and servicing*

10 Areas of uncertainty and requirements for further research

Further research and testing is required in most aspects of externally bonded FRP strengthening for metallic structures. It should be stressed that this report contains only design *guidance*. Experimental testing is recommended to validate the design procedures presented herein. Appropriate test programmes should also be undertaken as a part of major strengthening schemes.

Specific areas requiring further research are identified in the main body of the text and are summarised below.

Design

- The effect of differential thermal expansion on the adhesive joint
- delamination strength of wrought iron substrates
- laboratory tests on the application of FRP to riveted and bolted structures
- production of S–N curves for FRP strengthening systems
- experimental validation of elastic and fracture mechanics approaches to adhesive joint analysis for FRP-metal joints (including tapered plates)
- experimental validation of curtailment length requirements
- calibration of factors of safety for adhesion strength.

It is recommended that a simplified code of practice for the structural design and assessment of FRP strengthening schemes for metallic structures be developed. It is also desirable that software tools be developed to handle the more complex aspects of the analysis.

Spray-on FRP strengthening methods are under development at the University of British Columbia in Vancouver (Jones, 2001). Trials are being undertaken on concrete structures, but spray-on FRP is unlikely to be effective in strengthening metallic structures, where a much stiffer form of strengthening is required. Spray-on FRP results in a random orientation of short fibres, which has around 37 per cent of the stiffness of a unidirectional composite with the same fibre content. Spray-on FRP might be used in areas where high-stiffness strengthening is not required, possibly to strengthen connection details.

Installation

- The degree of surface preparation required to provide adequate bond
- the effect of existing substrate corrosion
- the structural consequences of workmanship tolerances
- the acceptable size and distribution of defects in the FRP and adhesive layer
- NDT techniques of inspection, in particular, of the adhesive bond
- installation procedures
- the durability effects of live loading applied to the structure during curing of the adhesive.

It is strongly recommended that an industry-wide certification scheme be established to ensure that the operatives and supervisors installing an FRP strengthening scheme have the appropriate training and experience.

In-service performance

- Prediction of long-term performance of FRP strengthening systems
- the prediction of substrate corrosion propagation beneath the adhesive layer
- the effect of induced currents in CFRP
- fire tests on FRP bonded to metallic structures
- the effect of high and low temperatures on FRP and adhesive strength and stiffness.

10.1 BIBLIOGRAPHY

Jones, N (2001). "Spray on strength". *New Scientist*, vol 172, no 2313, p 26

A1   Case histories

This appendix presents previous instances of externally bonded FRP strengthening applied to metallic structures. The information is split into two parts. Four representative case studies are given in detail:

- Hythe Bridge – prestressed HM CFRP strengthening of a cast iron bridge
- Boots Building – *in-situ*, pre-impregnated CFRP strengthening of a curved steel beam
- King Street Bridge – UHM CFRP strengthening and load-relief jacking of a cast iron bridge
- Tickford Bridge – strengthening of curved panels using wet lay-up CFRP.

Further examples of FRP strengthening are given in summary form in Table A1.1 to illustrate the range of strengthening projects that have previously been undertaken. Where available, references to further information have been given.

A1.1   TRANSFER OF PERMANENT STRESS BY PRESTRESSING

Hythe Bridge, Oxford, England

Constructed 1874, strengthened 2000

The first known application of the prestressed FRP technique for flexural upgrading cast iron bridges took place on the Hythe Bridge, Oxford, which was constructed in 1874. The objective of strengthening the bridge was to raise the capacity from Group 2 fire engines to 40 t vehicles. The description of the bridge strengthening has been given in Luke (2001a). The bridge was weak in mid-span bending, but was able to support the full 40t assessment in shear. The bridge consisted of two spans each of 7.8 m. The structure carries a busy city centre road above a tributary of the River Thames and a means of strengthening was required that did not cause traffic disruption.

A feasibility study was carried out, which showed that it was possible to strengthen the structure with either steel plate bonding, unstressed composite plates or stressed composite plates. Each of the three methods involved a degree of uncertainty because they extended previous practical limits in ways that threatened technical or economic viability. Steel plate bonding and unstressed composite plate bonding both presented significant difficulties, as the following comparison of the techniques indicates.

Steel plate bonding

- A thickness of 135 mm would have restricted head room and imposed high additional load
- such thickness would have required several layers of plates and was beyond previous experience
- difficulties in handling and fixing; drilling into cast iron was not advisable
- expensive to undertake, and the final system would require continuing maintenance.

Unstressed composite plate bonding

- Multiple layers of composite laminates would have been required, increasing labour and material costs beyond that normally associated with plate bonding

- unstressed strengthening would have required about 70 mm of HM CFRP (UHM CFRP was not available at the time, but would have required a thickness of around 35 mm)

- behaviour of such a high-build multilayer system is untested

- peel forces for such a thick system would require strapping, due to the absence of representative tests.

Stressed composite plate bonding

- No system for stressing composites in this situation was available

- stressing of cast iron could reveal weaknesses.

It was concluded that the stressed composite technique offered the most satisfactory solution. Four HM prestressed CFRP plates, 4.5 mm thick, were applied to the bottom flange of each beam, and were stressed to a total of 18 tonnes.

The stressing solution developed requires anchorages to be fixed to the extremities of the plate by bonding, friction or mechanical means (or a combination thereof). End tab plates are bonded to each end of the carbon fibre reinforced composite plates and provide a means of attaching jacking equipment and anchoring the plates when extended to the final working strain. The plates are stressed by a hydraulic jack, which reacts against a jacking frame temporarily fixed to the anchorage. The stressed plate is secured after extension by a shear pin that transfers load from a keyway in the end-tab to the anchorage. The composite plates are bonded to the beam by epoxy resin in addition to the end anchorages. The anchorage itself is surrounded by a protective casing and fully grouted. It was found that the stressing operation took about eight minutes for each strip of strengthening.

Further information

<http://www.balvac.co.uk/experience/projects/project_h4.html>

Luke (2001a), Luke (2001b), Luke (2001c)

Figure A1.1 *Hythe Bridge strengthening using prestressed CFRP plates (courtesy Mouchel – designer of the strengthening scheme)*

A1.2   STRENGTHENING USING PRE-IMPREGNATED COMPOSITES

Boots Building, Nottingham, England

A principal curved steel beam on the Boots Building in Nottingham was strengthened using a low-temperature moulding (LTM) advanced polymer composite material. Garden (2001) has described this strengthening project. The beam is curved in plan to connect two straight members around the corner of the building.

The purpose of the strengthening scheme was to restore the flexural and torsional capacity of the beam to above its original uncorroded level so that an anticipated increase in floor loading could be accommodated. The thickness of the composite strengthening layer, which comprised unidirectional, 0°/90° and ± 45° fibre orientation, was 2 mm. The unidirectional fibres were aligned along the direction of the beam's length for flexural strengthening. The 0°/90° fibre directions were used to resist shear and torsional loading, created due to the curvature of the beam. The prepreg composite layers were based upon glass and carbon fibres, using a low-temperature-curing epoxy resin.

Surface preparation for the steel adherend and installation procedure was undertaken as described in Section 3.2.1, supplemented by silica gel packs and enclosure in polythene, to keep the beam dry. To act as a bonding aid, an ambient-cured epoxy adhesive was painted on to the clean and dry steel surfaces.

The composite was installed using a cold-cure prepreg, under vacuum and cure temperature of 65°C.

Further information

Garden (2001)

Figure A1.2 *Strengthening a curved steel beam using pre-impregnated CFRP (courtesy Taylor Woodrow, Southall and ACG, Derbyshire)*

TRANSFER OF PERMANENT STRESS BY LOAD-RELIEF JACKING

King Street Bridge, Mold, Wales

Constructed 1870, strengthened 2000

King Street cast iron railway bridge was constructed in 1870 in Mold, Flintshire, and in 2000 was strengthened to allow 40 t vehicles to use it. Farmer and Smith (2001) have given a detailed discussion for the preservation of the cast iron structure; a brief description of the system is given here.

The construction of the bridge consists of brick arches spanning between six cast iron girders, which are supported on brick abutments. The bridge has a clear span of 8.93 m and consists of six cast iron girders skewed at an angle approximately 28°. Masonry jack arches, which span 1.99 m, are supported on the bottom flanges of the cast iron girder and have ties, which resist the lateral thrust.

The bridge carries the B544 road over the railway, which originally had two tracks. In the 1960s one track was removed and a propping system was installed beneath one half span. In the 1980s the borough council acquired the disused railway line to provide access between two car parks and took over responsibility for the bridge. In 1999 an assessment of the bridge capacity indicated that it was suitable only to carry vehicles up to 17 t gross weight. This was well short of the 40 t target, so a weight restriction was applied to the bridge while a strengthening scheme was designed.

Several designs for upgrading the bridge were proposed, and the one selected was to install temporary pre-loaded struts to support the deck, allowing the existing props to be removed and the strengthening materials applied before the temporary struts were removed. This system of load-relief jacking results in some of the permanent load being transferred in the FRP strengthening.

Ultra-high-modulus, unidirectional, carbon fibre and glass fibre laminate reinforcement was used as the strengthening material; this had a mean elastic modulus of elasticity of 360 GPa and a tensile strength of 1.1 GPa. The carbon fibres were selected for their high stiffness, and the glass fibres were used to give the laminates transverse strength and to prevent galvanic action from taking place.

On each of the four internal girders, two 170 mm × 33 mm-thick laminates were required to be positioned 153 mm apart, providing a sufficient gap for the heads of the hydraulic jacks to be positioned. The plates were tapered at the ends to a length of 2.5 m to reduce the stress concentrations at the ends of the plates to a minimum. The edge girders had a different cross-section to the internal girders. The thickness of the edge laminates were limited to 10 mm, also had a 153 mm space between them.

The surface of the cast iron was prepared by grit-blasting to achieve a specific texture and to remove any corrosion. Where necessary, localised repairs were made using a filled epoxy resin.

The laminates were manufactured with a small pre-camber of 10 mm. Because of the profile of girders, the thickness of the adhesive layer, applied to both the cast iron substrate and laminate surface, varied between 2 mm and 10 mm. During the curing cycle, clamps were used to apply a small force to the laminate and a weatherproof enclosure surrounded the plate to enable a temperature of at least 5°C to be maintained at all times. To ensure that the designed strains were not exceeded, and to monitor the performance of the bridge over several years, strain gauges were bonded at specific points to the plates and to the cast iron girders.

The advantages of using carbon fibre to upgrade this structure over other more conventional materials include:

- the durability of the material has been shown to be superior
- the loss of head room under the bridge due to the application of the layer of composite is minimal
- the erection procedure was simpler.

Further information

<http://www.tgp.co.uk/feature/frp4/kingstreet.html>

Farmer (2001)

Figure A1.3 *King Street Bridge strengthening using CFRP plates and load-relief jacking (courtesy Tony Gee and Partners – designer of the strengthening scheme)*

STRENGTHENING OF CURVED SURFACES

Tickford Bridge, Newport Pagnell, England

Constructed 1810, strengthened 1999

Tickford Bridge is believed to be the oldest operational cast iron highway bridge in the world. It was designed by Thomas Wilson and built over the River Ouzel in the summer of 1810 and is a scheduled ancient monument.

Following a structural assessment, a weight restriction of 3 tonnes was imposed on the bridge, causing disruption to local traffic. Strengthening was required to allow the weight restriction to be lifted. CFRP sheets were chosen by virtue of their strength, speed of installation (resulting in minimum disruption) and negligible impact on the aesthetics of the strengthened bridge. The scheme was found to be cheaper than traditional alternatives.

The selected system was a wet lay-up carbon fibre sheet strengthening system. This offered an advantage over other carbon fibre systems, which cannot be readily applied to curved surfaces. In critical areas, the curing temperature of the composite was elevated to between 50°C and 60°C. In addition, a continuous filament polyester drape veil was installed to provide insulation between the carbon fibre and the cast iron to avoid any possibility of galvanic corrosion. In total, 120 m^2 of carbon fibre prepreg sheet was applied in up to 14 layers and repainted, with a maximum thickness of 10 mm, having negligible visible affect on the bridge appearance. The work was completed within 10 weeks.

Further information

Lane and Ward (2000)

New Civil Engineer (2000)

Figure A1.4 *Strengthening of spandrel rings on Tickford Bridge (courtesy FaberMaunsell)*

Table A1.1 Selection of externally bonded FRP strengthening projects

Location	Date constructed/ date strengthened	Type of structure	Type of strengthening	Further information
Cast iron structures				
Shadwell station, London Underground	— / 2000	18 cruciform section cast iron struts in brick ventilation shaft	Up to 26 plies of UHM and HS CFRP, applied using vacuum infusion	Moy *et al* (2000), Leonard (2002)
Lamp standards, Tower Bridge, London	1894 / 2000	Cast iron lamp standards	Invisible repair of cracks that had formed due to thermal movements and traffic vibration	
Parapet railings, Ironbridge, Shropshire	1779 / 2000	Original cast iron railings on the world's first cast iron bridge	A minimal intervention repair to upgrade the parapet to modern crowd loading standards. 3.8 mm-thick CFRP plates were bonded to the inside faces of the parapet posts	*New Civil Engineer*, 14 Feb 2002
Bid Road Bridge, Hildenborough, Kent	1876 / 1999	Nine cast iron beams supporting brick jack arches	Tapered UHM CFRP plates were bonded to the beams to allow the bridge to carry 40 t vehicles	<http://www.concrete-repairs.co.uk/news/pr04_bridge0999.htm>
Bow Road Bridge, east London	1850 / 1999	Carried A11 over Docklands Light Railway. Cast iron beams supporting a combination of brick jack arches and steel plates	170 × 20 mm UHM CFRP plates were used to strengthen cast iron beams beneath the footways to support 40 t vehicles	<http://www.concrete-repairs.co.uk/news/pr07_carbon0300.htm>
Covered ways 12 & 58 Kelso Place, London Underground	1860 / 1999	Brick jack arches over cut-and-cover tunnels, supported by cast iron beams	CFRP plates were bonded to the underside of the beams to prevent overstressing while work was carried out on the foundations of the tunnel wall. The plates did not encroach on headroom requirements	Church and Silva (2002)
New Moss Road Bridge		Cast iron bridge over a canal	Unstressed UHM CFRP plates	Luke (2001a)

Location	Date constructed/ date strengthened	Type of structure	Type of strengthening	Further information
Redmile Canal Bridge		Cast iron bridge over a canal	Unstressed UHM CFRP plates	Luke (2001a)
Bridge EL31, Surrey Quays, London Underground		Cast iron beams and columns supporting brick jack arches and early riveted trough decking	HM CFRP strengthening applied to increase load capacity of bridge. Very limited access through ticket barriers	Church and Silva (2002)
Maunders Road bridge, Stoke on Trent	— / 2001	Cast iron beams supporting brick jack arches, carrying road over railway line	Tapered, UHM CFRP plates. Load-relief jacking used to transfer a proportion of the dead load into the CFRP and increase load capacity of bridge for heavy goods vehicles	

Wrought iron structures

No externally bonded FRP strengthening has been applied to wrought iron structures at the time of writing. The only reported use of composite to rehabilitate a wrought iron structure is a deck replacement scheme in Pennsylvania (Shenton et al, 2000)

Steel structures

Location	Date constructed/ date strengthened	Type of structure	Type of strengthening	Further information
Slattocks Canal Bridge	1936 / 2000	Longitudinal early riveted steel beams with reinforced concrete deck	HM CFRP plates applied to inner beams, allowing bridge to carry 40 t vehicles. Plastic capacity of beams mobilised in design	Luke (2001a) <http://www.balvac.co.uk/experience/projects/project_s2.html>
Underbridge D65A, Acton, London Underground	— / —	Early riveted plate girder steel bridge (I-section girders with timber deck)	UHM CFRP applied to cross-girders. Demonstration project addressing surface preparation requirements and curing of adhesive on a live railway structure	
Repair of corroded pipes	— / —	Sub-sea steel pipes (offshore infrastructure)	Wet lay-up HM CFRP patch around pipe	
Christina Creek Bridge 1-704, Newark, Delaware, USA	— /2001	Steel girder bridge with concrete deck on Interstate Highway 95	A demonstration project to investigate the fatigue durability and environmental resistance of the adhesive bond. CFRP was applied to a single girder	Mertz et al (2001) Miller et al (2001)
Steel blast wall, Mobil Beryl Platform	— / —	A steel blast wall, supported by universal columns	UHM composite applied to the columns to increase the overpressure capacity of the blast wall	
Type 42 destroyers, HMNB Portsmouth	1974–1983/ 1998–2001	Fatigue cracks were present in the food lifts of five Type 42 destroyers	Over 30 CFRP patches were applied to reinforced crack sites on five ships. Hand layup, resin infusion and prepreg application techniques were trialed	Dr T J Turton, QinetiQ, Farnborough GU14 0LX. 01252 395075, tjturton@qinetiq.com
Type 21 frigates, HMNB Portsmouth	1972–1975 / 1983–1984 (sold 1992–1993)	Fatigue cracks were present in the aluminium superstructure of Type 21 frigates	CFRP/epoxy patches were selected due to their ability to conform to the uneven deck surface	Dr T J Turton, QinetiQ, Farnborough GU14 0LX. 01252 395075, tjturton@qinetiq.com

A1.5   BIBLIOGRAPHY

Church, D G and Silva, T M D (2002). "Application of carbon fibre composites at covered ways 12 and 58 and bridge EL". In: *Proc ACIC 2002 – inaug int conf use of advanced composites in construction, 15–17 Apr, Univ of Southampton*. Thomas Telford, London

Farmer, N and Smith, I (2001). "King Street Railway Bridge – strengthening of cast iron girders with FRP composites". In: M C Forde (ed), *Proc 9th int conf struct faults and repair, 4–6 Jul, London*. Engineering Technics Press

Garden, H N (2001). "Use of composites in civil engineering infrastructure". *Reinforced plastics*, Jul/Aug, vol 45, no 7/8, pp 44–50

Lane, I R and Ward, J A (2000). "Restoring Britain's bridge heritage". Paper presented to Instn Civ Engrs (South Wales Association) Transport Engineering Group

Leonard, A R (2002). "The design of carbon fibre composite strengthening for cast iron struts at Shadwell Station vent shaft". In: *Proc ACIC 2002, inaug int conf use of advanced composites in construction, 15–17 Apr, Univ of Southampton*. Thomas Telford, London

Luke, S (2001a). "Strengthening of structures with carbon fibre plates – case histories for Hythe Bridge, Oxford and Qafco Prill Tower, Qatar". In: *Proc NGCC 1st ann conf & AGM, Composites in construction; through life performance, 30–31 Oct, Watford*

Luke, S (2001b). "Strengthening of existing structures using advanced composite materials". NGCC paper presented at IStructE, 18 Sep 2001

Luke, S (2001c). "The use of carbon fibre plates for the strengthening of two metallic bridges of an historic nature in the UK". In J G Teng (ed), *Proc int conf FRP composites in construction (CICE 2001), 12–15 Dec, Hong Kong*, pp 975–983

Mertz, D R, Gillespie, J W, Chajes, M J and Sabol S A, (2001). *The rehabilitation of steel bridge girders using advanced composite materials*. IDEA program final report, con no NCHRP-98-ID051, Transportation Research Board, National Research Council

Miller, T C, Chajes, M J, Mertz, D R and Hastings, J N (2001). "Strengthening of a steel bridge girder using CFRP plates". In: *Proc New York City bridge conf, 29–30 Oct, NY*

Moy, S S J, Barnes, F, Moriarty, J, Dier, A F, Kenchington, A, Iverson B (2000). "Structural upgrade and life extension of cast iron struts and beams using carbon fibre reinforced composites". In: A G Gibson (ed), *Proc 8th int conf fibre reinf composites, FRC 2000 – composites for the millennium, 13–15 Sep, Univ Newcastle-upon-Tyne*

New Civil Engineer (2000). "Historic Bridge Award 2000 winners". *New civ engr*, no 1378, 23 Nov, p 26

New Civil Engineer (2002). "Custodian of the lists. Joining the cast". *New civ engr*, no 1429, 14 Feb, p 25

Shenton, H W, Chajes, M J, Finch, W W, Hemphill, S and Craig, R (2000). "Performance of a historic 19th century wrought iron through-truss bridge rehabilitated using advanced composites". In: *Proc ASCE advanced tech in structural engg: structures congress 2000, 8–10 May, Philadelphia*. ASCE, Reston, VA

A2   Manufacturing methods for composite reinforcing fibres

A2.1   CARBON FIBRE

Carbon fibres are manufactured by controlled pyrolysis and crystallisation of an organic precursor. The carbon crystallites are preferably orientated along the fibre length. As the manufacturing temperature increases, the size and orientation of the crystallites improves, the crystallinity of the fibre increases and consequently the degree of amorphous material decreases. This results in an exponential increase in the modulus of elasticity with temperature. However, the tensile strength increases in value as the temperature rises to a maximum of about 1600°C and then falls to a constant value as the temperature continues to rise to give the required value of modulus.

The three types of carbon fibre thus depend upon the heat treatment used during their manufacture.

To produce carbon fibres successfully using high processing temperatures, it is essential that the precursor carbonises and does not melt, allowing it to be converted into fibres. There are three precursors used to manufacture carbon fibres.

1. *Rayon precursors* are derived from cellulose materials and were one of the earliest precursors used to make carbon fibres, but only about 25 per cent of the original fibre mass remains after carbonisation.

2. *Polyacrylonitrile (PAN) precursors* are the basis for the majority of carbon fibres that are commercially available and about 50 per cent of the original fibre mass remains after carbonisation. Carbon fibres made from PAN precursors have higher tensile strengths than other precursor-based fibres. Currently all commercial production of PAN precursor carbon fibres is by spinning and the cross-section produced is round. A new production system for an acrylic-based precursor fibre involves a melt-assisted extrusion as a part of the spinning operation. This method allows cross-sections of rectangular, I-type and X-type to be produced, permitting closer fibre packing in the composite.

3. *Pitch precursors* are based upon petroleum, asphalt, coal tar and PVC, and are relatively low in cost and high in carbon yield, although the fibre manufactured from batch to batch is generally less uniform than with PAN precursors.

High-strength (HS) and high-modulus (HM) fibres are generally produced from polyacrylonitrile precursors, and ultra-high modulus (UHM) fibres are produced from a pitch precursor. UHM fibres are very brittle, so care must be taken when handling them.

A2.2   GLASS FIBRE

Glass fibres are generally manufactured by the direct melt process in which fine filaments of diameters 3–24 μm in commercial production are produced by continuous and rapid drawing from the melt. During the production stage, strands, each consisting of 200 individual filaments, are formed. Glass filaments are highly abrasive to each other, so to minimise abrasion-related degradation of the glass fibre, surface treatment or sizing is applied before the fibres are gathered into strands and wound onto a drum. The size also binds the filaments together and acts as a coupling agent to the matrix during impregnation. The four main functions of the sizing operation are:

- to facilitate the manufacturing technique
- to reduce the abrasive effect of the filaments against one another
- to reduce damage to the fibres during mechanical handling
- to provide a chemical link between the glass fibre and the polymer.

Glass fibres are silica-based glass compounds that contain several metal oxides which can be tailored to create different types of glass. The main oxide is silica in the form of silica sand; the other oxides such as calcium, sodium and aluminium are incorporated to reduce the melting temperature and impede crystallisation. The important typical mechanical properties are given in Table 3.1.

The most important grades of glass are listed below.

E-glass has a low alkali content of the order of 2 per cent. It is used for general-purpose structural applications and is the major one used in the construction industry; it also has good heat and electrical resistance.

S-glass is a stronger (typically 40 per cent greater strength at room temperature) and stiffer fibre with a greater corrosion resistance than the E-glass fibre. It has good heat resistance. The S-2-glass has the same glass composition as S-glass but differs in its coating. S-2-glass has good resistance to hydrochloric, nitric, sulphuric and other acids.

E-CR-glass has good resistance to acids and bases and has chemical stability in chemically corrosive environments.

R-glass has a higher tensile strength and tensile modulus and greater resistance to fatigue, ageing and temperature corrosion than does E-glass.

AR-glass is alkali-resistant and is used as the reinforcement for glass fibre reinforcement cement (GRFC) When glass fibres are used to reinforce cement, degradation in strength and toughness occurs when it is exposed to outdoor weathering, especially in humid conditions. This process can also take place with AR-glass, albeit at a much slower rate.

A2.3 ARAMID FIBRE

The aromatic fibre (the aramid fibre) is typically produced by an extrusion and spinning process. A solution of the polymer in a suitable solvent at a temperature of between -50°C and -80°C is extruded into a hot cylinder, which is at a temperature of 200°C; this causes the solvent to evaporate and the resulting fibre is wound on to a bobbin. The fibre is then stretched and drawn to increase its strength and stiffness. The molecular chains are aligned and made rigid by means of aromatic rings linked by hydrogen bridges. Two grades of stiffness are generally available; one has a modulus of elasticity in the range of 60 GPa and the other has a modulus of elasticity of 130 GPa. The higher-modulus fibre is the one that is used in polymer composites in construction.

The structure of the aramid fibres is anisotropic and gives higher strength and modulus values in its longitudinal direction compared with its transverse direction. Aramid is resistant to fatigue, both static and dynamic. It is elastic in tension but it behaves non-linearly in compression and in addition has a ductile compressive characteristic. The fibre possesses good toughness and damage tolerance properties.

A4   Fatigue in composite materials

Figure A4.1 shows a possible sequence of fatigue damage when a unidirectional composite is loaded parallel to its fibres.

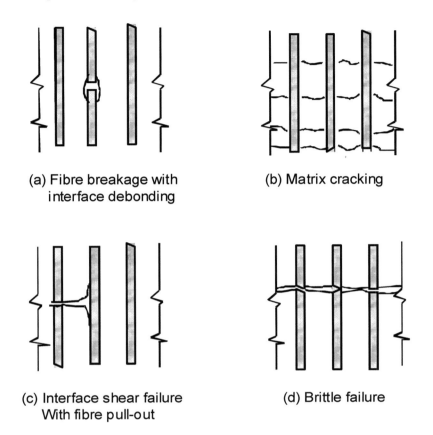

(a) Fibre breakage with
interface debonding

(b) Matrix cracking

(c) Interface shear failure
With fibre pull-out

(d) Brittle failure

Figure A4.1 *Damage mechanism in a unidirectional composite under load parallel to the fibres*

In homogeneous metals, crack propagation is perpendicular to the cyclic load axis. If a composite is subjected to loading in a direction that is not parallel to its fibres, a highly "diffused damage" zone is created where cracks are multiplied. Metals are more sensitive to tensile loading than compressive loading, whereas the converse is true for composites.

A second difference between fatigue in metallic and fatigue in composite materials is that the reduction in stiffness and residual strength begins very early in the fatigue life and reaches significant magnitudes long before the component breaks in the composite. Figure A4.2 compares damage accumulation in homogeneous monolithic materials with that in laminated composites.

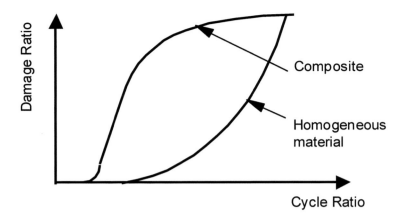

Figure A4.2 *Damage accumulation in homogeneous and laminated materials*

Fatigue in FRP materials is influenced by the following parameters.

(a) *The fibre and matrix materials*

Carbon fibre reinforced composites have excellent fatigue performance compared with metallic and other composite materials. If the failure strain of the matrix exceeds that of the fibre, fibre fracture will dominate fatigue failure (such as in high-modulus CFRPs). Otherwise, interfacial shear and matrix cracking will predominate at low cycles, with fibre fracture only occurring late in the fatigue life of the composite. A toughened epoxy matrix has a higher static strength than a standard epoxy, but has a reduced fatigue performance.

(b) *Interface effects*

The interface between the fibre and the matrix plays an important role in the fatigue behaviour of the FRP composites. A strong interface delays the occurrence of fibre ridging and longitudinal matrix cracking, thus improving the fatigue performance of the unidirectional FRP composite material.

(c) *Frequency effects*

Fatigue testing is often undertaken at high frequencies, in contrast to the low-frequency fatigue loading experienced by typical structures. Internal heat build-up during high-frequency testing can reduce the fatigue life of an FRP test specimen (Demers, 1998 and Rotem, 1993). However, in some fatigue frequency investigations it has been found that higher frequencies resulted in a longer fatigue length (Mandell, 1982).

(d) *Manufacturing process*

The fatigue behaviour of FRP composites is highly dependent upon the microstructure of the FRP, which depends upon the manufacturing technique.

A4.1 BIBLIOGRAPHY

Adams, R D, Comyn, A and Wake, W C (1997). *Structural adhesive joints in engineering.* 2nd edn, Chapman and Hall

Bussell, M N (1997). *Appraisal of existing iron and steel structures.* P138, Steel Construction Institute, Ascot

Concrete Society (2000). *Design guidance for strengthening concrete structures using fibre composite materials.* Technical Report 55, Concrete Society, Crowthorne

Concrete Society (2003). *Strengthening concrete structures with fibre composite materials: acceptance, inspection and monitoring.* Technical Report 57, Concrete Society, Crowthorne

Cripps, A, ed (2001). *Fibre-reinforced polymer composites in construction.* C564, CIRIA, London

fib (2001). *Externally bonded FRP reinforcement for RC structures.* fib bulletin 14, Fédération internationale du béton, Lausanne

IStructE (1996). *Appraisal of existing structures.* 2nd edn, SETO, London

IStructE (1999). *A guide to the structural use of adhesives.* SETO, London

Mandell, J. (1982). "Fatigue behaviour in fibre resin composites". In: G Pritchard (ed), *Developments in reinforced plastics.* Kluwer, Dordrecht, vol 2, pp 67–107

Moy, S S J, ed (2001b). *FRP composites – life extension and strengthening of metallic structures.* ICE design and practice guide, Thomas Telford, London

A5 ELASTIC ANALYSIS OF THE ADHESIVE STRESS DISTRIBUTION FOR A PLATE OF CONSTANT THICKNESS

As described in Section 5.3, a strain discontinuity in the substrate, adhesive or FRP strengthening material leads to a stress concentration in the adhesive. The strain discontinuity could be due to (Figure 5.8):

- the end of a piece of FRP strengthening (FRP not continuous)
- a crack in the metallic substrate (substrate not continuous)
- a defect in the adhesive (adhesive not continuous).

The elastic adhesive joint analysis in this appendix allows the stress distribution in the adhesive layer to be determined, and thus the capacity of the adhesive joint checked with respect to the strength of the adhesive. The analysis has been checked as far as possible by a desk study, but experimental work has not been carried out to validate the analysis.

A number of closed form solutions have been developed for adhesion (for example, see the bibliography at the end of this appendix). The analysis presented here is similar to previous approaches. Additional information on its derivation will be published in Cadei and Stratford (2004).

A5.1 LACK OF FIT ACROSS THE ADHESIVE JOINT

If the substrate and the FRP *were not bonded together by the adhesive*, there would be a lack of fit between the substrate and the FRP.

The lack of fit at the adhesive interface is due to:

- prestress in the FRP strengthening material
- differential thermal expansion
- the additional loads applied to the metallic section after the strengthening is bonded to the metallic substrate.

These loads are defined by:

$(T\text{-}T_0)$ the temperature change since the adhesive set

N_s the axial force resulting from loads applied to the strengthened member *after the adhesive set* (tension positive)

M_s the bending moment resulting from loads applied to the strengthened member *after the adhesive set*, applied at the neutral axis level of the metallic section (hogging positive)

N_{f0} the initial force in the FRP at the time of strengthening, due to prestress (tension positive)

M_{f0} the initial bending moment in the FRP (hogging positive). This is usually zero, but should be used if the supplied FRP plate is initially curved.

If the adhesive bond were released, the applied loads (N_s, M_s) would be carried by the metallic section alone, giving a strain (ε_{sR}) and curvature (ψ_{sR}) in the metallic substrate of:

$$\varepsilon_{sR} = \frac{N_s}{E_s A_s} - \psi_{sR}\, y_s + \alpha_s \left(T_s - T_{s0}\right), \text{ at the adhesive interface} \qquad (A5.1)$$

$$\psi_{sR} = \frac{M_s}{E_s I_s} \qquad (A5.2)$$

After the adhesive has cured, any prestress is transferred to the FRP, and this prestress is maintained only by the adhesive bond. Should the adhesive bond be released, the FRP would contract back to its original length. The strain (ε_{fR}) and curvature (ψ_{fR}) in the FRP in the absence of bond would be:

$$\varepsilon_{fR} = -\frac{N_{f0}}{E_f A_f} + \psi_{fR} y_f + \alpha_f \left(T_f - T_{f0}\right), \text{ at the adhesive interface} \qquad (A5.3)$$

$$\psi_{fR} = -\frac{M_{f0}}{E_f I_f} \qquad (A5.4)$$

(The subscript "R" indicates a strain in the FRP or substrate if the lack of fit was released. Strains are positive in tension, curvatures are positive in hogging.)

The lack of fit across the adhesive joint is due to the difference in strains and curvatures between the FRP and the substrate. Both of these can vary along the beam (in x).

The lack-of-fit axial strain,

$$\Delta\varepsilon_{fs}(x) = \varepsilon_{fR} - \varepsilon_{sR} \qquad (A5.5)$$

The lack-of-fit axial curvature,

$$\Delta\psi_{fs}(x) = \psi_{fR} - \psi_{sR} \qquad (A5.6)$$

Note that the analysis assumes a plane-section strain distribution in the metallic section; however, in the vicinity of the crack a plane-section assumption will not be strictly true.

A5.2 GENERAL GOVERNING EQUATIONS

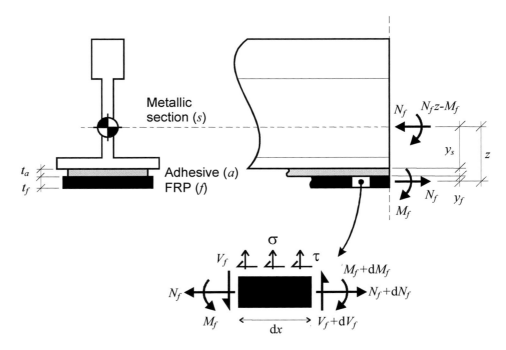

Figure A5.1 *Equilibrium of the beam, the FRP composite strengthening and the adhesive*

The strengthened section is defined by the following parameters (see also Figure A5.1):

t_a thickness of adhesive

b_a width of adhesive

E_a Young's Modulus of adhesive

G_a shear modulus of adhesive

t_f thickness of FRP strengthening material

E_f Young's Modulus of FRP

I_f second moment of area of FRP

A_f cross-sectional area of FRP

y_f distance from the centroid of the FRP strengthening to the adhesive interface (usually equal to $t_f/2$)

E_s Young's Modulus of the metallic member

I_s second moment of area of the (unstrengthened) metallic member

A_s cross-sectional area of the metallic member

y_s distance from the centroid of the beam to the adhesive interface. (If the FRP is applied to the soffit of the beam, this will be equal to y_{g0})

z lever arm between centroid of the *un*strengthened metallic member and the centroid of the FRP material

x position along the beam.

τ shear stress in adhesive, at centre of adhesive joint

σ normal, peel stress in adhesive, at centre of adhesive joint

N_f axial force in the FRP strengthening material

M_f bending moment in the FRP strengthening material.

Equilibrium of a short length of the FRP can be used to write the shear and peel stresses in the adhesive, in terms of the axial force (N_f) and moment (M_f) in the FRP strengthening material (Figure A5.1):

$$\tau = \frac{1}{b_a}\frac{dN_f}{dx}$$ (A5.7)

and

$$\sigma = -\frac{1}{b_a}\left(\frac{\mathrm{d}^2 M_f^*}{\mathrm{d}x^2}\right)$$ (A5.8)

where

$$M_f^* = M_f - y_f N_f$$ (A5.9)

is the transformed plate bending moment about the interface between the strengthening and the adhesive.

The substrate and the FRP strengthening material are bonded together by the adhesive. Thus, there can be no lack-of-fit between the FRP and the substrate, and compatibility must be satisfied, as shown in Figure A5.2.

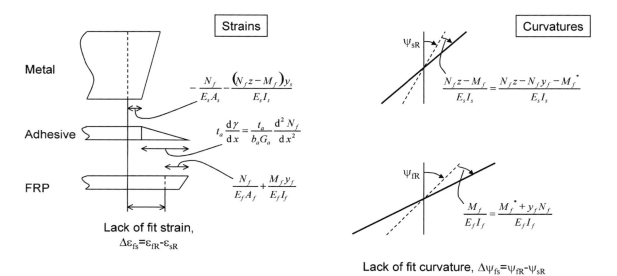

Figure A5.2 *Compatibility requirements across the adhesive joint*

Compatibility requirements in the adhesive layer govern the variation in axial force and bending moment in the FRP (N_f, M_f) along the beam required to ensure there is no lack of fit:

$$-\left(\frac{t_a}{G_a b_a}\right)\frac{d^2 N_f}{dx^2} + \left(\frac{1}{E_f A_f} + \frac{1}{E_s A_s} + \frac{z y_s}{E_s I_s}\right)N_f + \left(\frac{y_f}{E_f I_f} - \frac{y_s}{E_s I_s}\right)M_f = -\Delta\varepsilon_{fs}(x) \qquad \text{(A5.10)}$$

and

$$\left(\frac{t_a}{E_a b_a}\right)\frac{d^4 M_f^*}{dx^4} + \left(\frac{1}{E_f I_f} + \frac{1}{E_s I_s}\right)M_f^* - \left(\frac{z-y_f}{E_s I_s} - \frac{y_f}{E_f I_f}\right)N_f = -\Delta\psi_{fs}(x) \qquad \text{(A5.11)}$$

A5.3   ADHESIVE SHEAR STRESS DISTRIBUTION

The two compatibility equations are coupled. To simplify the solution, it can be assumed as a first approximation that the curvatures of the plate and beam are equal:

$$\frac{M_f}{E_f I_f} = \frac{N_f z - M_f}{E_s I_s} \qquad \text{(A5.12)}$$

The shear compatibility equation (A5.10) then becomes (by substitution for M_f):

$$-f_1 \frac{d^2 N_f}{dx^2} + f_2 N_f = -\Delta\varepsilon_{fs}(x) \qquad \text{(A5.13)}$$

where

$$f_1 = \left(\frac{t_a}{G_a b_a}\right) \qquad \text{(A5.14)}$$

is the flexibility of the adhesive layer, and

$$f_2 = \left(\frac{1}{E_f A_f} + \frac{1}{E_s A_s} + \frac{z(z - t_a)}{E_s I_s + E_f I_f}\right) \qquad \text{(A5.15)}$$

is the relative flexibility of the FRP and beam.

Solving this equation gives:

$$N_f = N_{PS} + C_1 e^{-\lambda x} \qquad \text{(A5.16)}$$

where:

$$\lambda = \sqrt{\frac{f_2}{f_1}} \qquad \text{(A5.17)}$$

is a measure of the relative flexibility of the beam, FRP and adhesive.

N_{PS} is a particular solution that, for a constant or linear variation in the lack of fit along the beam, is:

$$N_{PS} = -\frac{\Delta\varepsilon_{fs}(x)}{f_2}$$ (A5.18)

The adhesive shear stress distribution is found from the axial force distribution using Equation A5.7:

$$\tau = \frac{1}{b_a}\left(\frac{dN_{PS}}{dx} - C_1 \lambda e^{-\lambda x}\right)$$ (A5.19)

The constant C_1 is found by applying a suitable boundary condition, which depends upon the position of the strain discontinuity that is of interest. As discussed in Section 5.3, the strain discontinuity could be due to (a) a defect in the adhesive bond line, (b) a crack in the substrate, or (c) the end of the FRP strengthening.

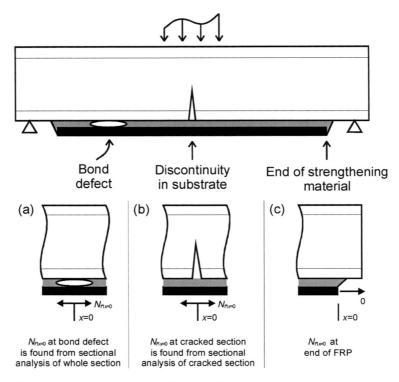

Figure A5.3 *Boundary conditions at the end of a plate and at a cracked section*

To determine a solution in the general case, let $x = 0$ correspond to the position of the strain discontinuity, and the force in the FRP strengthening member to be $N_f\big|_{x=0}$ at that position. The values of $N_f\big|_{x=0}$ for different strain discontinuities are indicated in Figure A5.3. For this boundary condition:

$$C_1 = N_f\big|_{x=0} - N_{PS}\big|_{x=0}$$ (A5.20)

where $N_{PS}\big|_{x=0}$ is the value of the particular solution at $x = 0$.

The distribution of shear stress near the end of a strengthening plate is plotted in Figure 5.9.

A5.3.1 Maximum shear stress

Of most importance for design is the maximum shear stress, which occurs at $x = 0$:

$$\tau_{max} = \frac{1}{b_a} \left[\left. \frac{dN_{PS}}{dx} \right|_{x=0} - \lambda C_1 \right] \tag{A5.21}$$

When Equation A5.18 holds, this becomes:

$$\tau_{max} = -\frac{1}{b_a} \left[\frac{1}{f_2} \left. \frac{d(\Delta\varepsilon_{fs})}{dx} \right|_{x=0} + \lambda \left(\left. N_f \right|_{x=0} + \frac{\left. \Delta\varepsilon_{fs} \right|_{x=0}}{f_2} \right) \right] \tag{A5.22}$$

From Equations A5.1 to A5.5, if the prestress and temperature do not vary along the beam, the only variation in $\Delta\varepsilon_{fs}$ with x will usually be the change in moment (M_s). Thus $d(\Delta\varepsilon_{fs})/dx$ is due to flexural shear stresses, which are usually negligible compared with the shear stress concentration in the adhesive.

$$\tau_{max} \cong -\frac{\left. \Delta\varepsilon_{fs} \right|_{x=0}}{b_a \sqrt{f_1 f_2}} - \frac{\lambda \left. N_f \right|_{x=0}}{b_a} \tag{A5.23}$$

This equation is used in Section 5.3.1 to predict the maximum shear stress (Equation 5.47).

A5.4 NORMAL PEEL STRESSES

Having obtained a solution for N_f (Equation A5.16) using the uncoupled shear compatibility equation, N_f may be treated as a known term in the peel compatibility equation, A5.11, which becomes:

$$a_1 \frac{d^4 M_f^*}{dx^4} + a_2 M_f^* = a_3 N_f - \Delta\psi_{fs}(x) \tag{A5.24}$$

where

$$a_1 = \frac{t_a}{E_a b_a} \tag{A5.25}$$

$$a_2 = \frac{1}{E_f I_f} + \frac{1}{E_s I_s} \tag{A5.26}$$

$$a_3 = \frac{z - y_f}{E_s I_s} - \frac{y_f}{E_f I_f} \tag{A5.27}$$

The solution of this equation gives the variation in the bending moment in the strengthening:

$$M_f^* = M_{PS}^* + C_2 e^{-\beta x} \cos \beta x + C_3 e^{-\beta x} \sin \beta x \tag{A5.28}$$

where:

$$\beta = \left(\frac{a_2}{4a_1}\right)^{0.25}$$

(A5.29)

The particular solution moment $M_{PS}{}^*$ includes the lack of fit terms, and the resultant moment from the axial force in the strengthening. Assuming that $\Delta\psi_{fs}$ and $\Delta\varepsilon_{fs}$ contain cubic or lower order terms in x:

$$M_{PS}^* = C_4(x) + C_5 e^{-\lambda x}$$

(A5.30)

where C_4 and C_5 are functions of the previously determined solution for N_f. They are:

$$C_4 = -\frac{1}{a_2}\left(\Delta\psi_{fs}(x) + a_3 \frac{\Delta\varepsilon_{fs}(x)}{f_2}\right)$$

(A5.31)

$$C_5 = \frac{a_3 C_1}{a_1 \lambda^4 + a_2}$$

(A5.32)

The peel stress is found from Equations A5.8 and A5.28:

$$\sigma = -\frac{1}{b_a}\left(\frac{d^2 M_{PS}^*}{dx^2} + 2C_2 \beta^2 e^{-\beta x} \sin\beta x - 2C_3 \beta^2 e^{-\beta x} \cos\beta x\right)$$

(A5.33)

The moment ($M_f{}^*$) and shear force ($dM_f{}^*/dx$) in the strengthening are known at the strain discontinuity ($x = 0$), and are used to give the boundary conditions constants C_2 and C_3. At the end of a plate, both the moment and shear force are zero, hence:

$$C_2 = -M_{PS}^*\big|_{x=0}$$

(A5.34)

$$C_3 = -\frac{1}{\beta}\left[\frac{dM_{PS}^*}{dx}\bigg|_{x=0} - \beta C_2\right]$$

(A5.35)

A5.4.1   Maximum peel stress

The maximum peel stress (at $x = 0$) is:

$$\sigma_{max} = -\frac{1}{b_a}\left[\frac{d^2 M_{PS}^*}{dx^2}\bigg|_{x=0} - 2C_3 \beta^2\right]$$

(A5.36)

If the prestress and temperature do not vary along the beam, the only variation in $\Delta\varepsilon_{fs}$ and $\Delta\psi_{fs}$ with x will usually be the change in moment (M_s) (see Equations A9.1 to A9.6). In most practical cases, d^2C_4/dx^2 is negligible compared with $\lambda^2 C_5$, and the expression for maximum peel stress simplifies to:

$$\sigma_{max} = -\frac{1}{b_a}\left[C_5 \lambda^2 - 2C_3 \beta^2\right]$$

(A5.37)

This is the equation for σ_{max} given in Section 5.3.1 as Equation 5.57.

A5.5   COMMENTS

It should be noted that the above analysis is based upon a number of assumptions. In particular, the adhesive is assumed to be linear-elastic, which is unlikely to be strictly true. Equilibrium considerations require the shear stress in the adhesive be zero at the very end of the strengthening. As shown by Figure 5.8, this boundary condition is not satisfied; the analysis slightly overestimates the peak shear stress.

A5.6   BIBLIOGRAPHY

Albat, A M and Romilly, D P (1999). "A direct linear-elastic analysis of double symmetric bonded joints and reinforcements". *Composite science and technology*, vol 59, no 7, pp 1127–1137

Cadei, J M C and Stratford, T J (2004). "Elastic analysis of adhesion stresses between a beam and a bonded strengthening plate". To be presented at ACIC-2004

Denton, S R (2001). "Analysis of stresses developed in FRP plated beams due to thermal effects". In: J G Teng (ed), *Proc int conf FRP composites in civ engg (CICE 2001), 12–15 Dec, Hong Kong*, pp 527–536

Frost, S, Lee, R J and Thompson, V K (2003). "Structural integrity of beams strengthened with FRP plates – analysis of the adhesive layer". In: *Proc struct faults and repair 2003, London*

Miller, T C, Chajes, M J, Mertz, D R and Hastings, J N (2001). "Strengthening of a steel bridge girder using CFRP plates". In: *Proc New York City bridge conf, 29–30 Oct, New York*

Roberts, T M (1989). "Approximate analysis of shear and normal stress concentrations in the adhesive layer of plated RC beams". *The structural engineer*, vol 10, no 2, pp 229–233

Smith, S T and Teng, J G (2001). "Interfacial stresses in plated beams". *Engineering structures*, vol 23, pp 857–871

A6   Analysis of the adhesive stress distribution in lap-shear specimens

Lap-shear test results are usually reported as an average shear strength in the adhesive. The adhesive stress, however, varies along the joint (Figure 5.11). It is the peak adhesive stress that is of importance, and which must be used in the design of an adhesive joint.

The following expressions allow the peak adhesive stress to be calculated from the average shear stress (expressed as the applied load, P, at failure). Both single lap and double lap-shear tests are covered. They are taken from Eurocomp (Clarke, 1996) Section 5.3.5.5. The expressions are lengthy, but they are of closed-form and essentially straightforward. It is suggested that a spreadsheet be used to aid multiple calculations.

There are several lap-shear test standards, each of which uses specimens with different geometries. The equations in this appendix can be applied to any lap-shear test in which the two pieces of adherend are the same. Section 7.1.1 discusses lap-shear test requirements for strengthening metallic structures using FRP and the recommended test method.

A6.1   SINGLE LAP-SHEAR TEST SPECIMEN

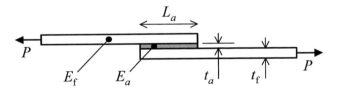

Figure A6.1   *A single lap-shear specimen*

For a single lap-shear specimen, the peak shear stress can be found using the following expressions (derived by Goland and Reissner, 1944):

The peak shear stress is at the joint end and is:

$$\tau_{max} = \begin{cases} \dfrac{P}{8b_a c_a}\left[\dfrac{\beta_s c_a}{t_f}(1+3k)\coth\left(\dfrac{\beta_s c_a}{t_f}\right)+3(1-k)\right] & \text{for } \dfrac{\beta_s c_a}{t_f}<25 \\[4mm] \dfrac{P}{8b_a t_f}(1+3k)\sqrt{8\dfrac{G_a t_f}{E_f t_a}} & \text{for } \dfrac{\beta_s c_a}{t_f}\geq 25 \end{cases} \qquad (A6.1)$$

The peak peel stress is at the joint end, and is:

$$\sigma_{max} = \begin{cases} \dfrac{P}{b_a t_f}\left(\dfrac{t_f}{c_a}\right)^2\left[\lambda_s^2\dfrac{k}{2}\dfrac{\sinh(2\lambda_s)-\sin(2\lambda_s)}{\sinh(2\lambda_s)+\sin(2\lambda_s)}-\lambda_s k'\dfrac{\cosh(2\lambda_s)+\cos(2\lambda_s)}{\sinh(2\lambda_s)+\sin(2\lambda_s)}\right] & \text{for } \lambda_s<2.5 \\[4mm] \dfrac{P}{b_a t_f}\left(\dfrac{k}{2}+k'\dfrac{t_f}{c_a}\right)\sqrt{6\dfrac{E_a\,t_f}{E_f\,t_a}} & \text{for } \lambda_s\geq 2.5 \end{cases} \qquad (A6.2)$$

where:

$$c_a = \frac{L_a}{2} \tag{A6.3}$$

$$\beta_s = \sqrt{8 \frac{G_a\, t_f}{E_f\, t_a}} \tag{A6.4}$$

$$\lambda_s = \frac{c_a}{t_f}\left(\frac{6E_a t_f}{E_f t_a}\right)^{\frac{1}{4}} \tag{A6.5}$$

$$k = \frac{\cosh(u_4 c_a)\sinh(u_3 L_a)}{\sinh(u_3 L_a)\cosh(u_4 c_a) + 2\sqrt{2}\,\cosh(u_3 L_a)\sinh(u_4 c_a)} \tag{A6.6}$$

$$k' = k\frac{c_a}{t_f}\left(3\!\left(1 - v_f^{\,2}\right)\frac{P}{b_a t_f E_f}\right)^{\frac{1}{2}} \tag{A6.7}$$

$$u_3 = 2\sqrt{2}u_4 \tag{A6.8}$$

$$u_4 = \frac{1}{\sqrt{2}t_f}\sqrt{3\!\left(1 - v_f^{\,2}\right)\frac{P}{b_a t_f E_f}} \tag{A6.9}$$

P	is the net load applied to the specimen
L_a	is the length of the adhesive overlap
b_a	is the width of the adhesive
t_f	is the adherend thickness
E	is the adherend tensile modulus
G_a	is the adhesive shear modulus
t_a	is the adhesive thickness
E_a	is the adhesive tensile modulus
v_f	is the adherend Poisson's ratio

Once the peak shear (τ_{max}) and peel stress (σ_{max}) have been determined, the strength of the adhesive ($\overline{\sigma}$) is found from (see Equation 5.38):

$$\overline{\sigma} = \frac{\sigma_{max}}{2} + \sqrt{\left(\frac{\sigma_{max}}{2}\right)^2 + \tau_{max}^{\,2}} \tag{A6.10}$$

A6.2   DOUBLE LAP-SHEAR TEST SPECIMEN

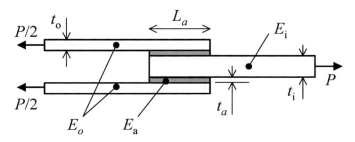

Figure A6.2 *A double lap-shear specimen*

For a double lap-shear specimen, the peak shear stress can be found using the following expressions (based on Goland and Reissner's theory for single lap-shear joints, but modified by Hart-Smith, 1973):

The peak shear stress is at the joint end, and is:

$$\tau_{max} = \frac{\lambda_d P}{4 b_a} \left[\frac{\cosh(\lambda_d c_a)}{\sinh(\lambda_d c_a)} + \Omega \frac{\sinh(\lambda_d c_a)}{\cosh(\lambda_d c_a)} \right] \tag{A6.11}$$

The peak peel stress is at the joint end, and is:

$$\sigma_{max} = \tau_{max} \left(\frac{3 E_a'\left(1 - v_o^2\right) t_o}{E_o t_a} \right)^{\frac{1}{4}} \tag{A6.12}$$

where:

$$c_a = \frac{L_a}{2} \tag{A6.13}$$

$$\lambda_d{}^2 = \frac{G_a}{t_a} \left(\frac{1}{E_o t_o} + \frac{2}{E_i t_i} \right) \tag{A6.14}$$

$$\chi = \frac{E_i t_i}{2 E_o t_o} \tag{A6.15}$$

$$\Omega = \max\left(\frac{1 - \chi}{1 + \chi}, \quad \frac{\chi - 1}{1 + \chi} \right) \tag{A6.16}$$

$$E'_a = \left[\frac{(1 - v_a)}{(1 + v_a)(1 - 2v_a)} \right] E_a \tag{A6.17}$$

P	is the net load applied to the specimen
b_a	is the width of the adhesive
t_o	is the thickness of the outer two adherends
t_i	is the thickness of the inner adherend
E_o	is the stiffness of the outer two adherends
E_i	is the stiffness of the inner adherend
v_o	is the Poisson's ratio of the outer adherend
t_a	is the thickness of the adhesive
E_a	is the stiffness of the inner adherend
G_a	is the adhesive shear modulus
v_a	is the Poisson's ratio of the adhesive

The maximum shear and maximum peel stresses are not coincident, but it is conservative to assume that they are. As for the single lap-shear test, the strength of the adhesive ($\overline{\sigma}$) is found from Equation A6.10:

$$\overline{\sigma} = \frac{\sigma_{max}}{2} + \sqrt{\left(\frac{\sigma_{max}}{2}\right)^2 + \tau_{max}^{\;2}} \qquad\qquad (A6.10bis)$$

A6.3 BIBLIOGRAPHY

Clarke, J L, ed (1996). *Structural design of polymer composites. Eurocomp design code and handbook*. E & FN Spon, London

Goland, M and Reissner, E (1944). "The stresses in cemented joints". *J app mechanics, trans ASME*, no A17

Hart-Smith, L J (1973). *Adhesive-bonded double-lap joints*. Contractor Report CR-112235, NASA, Langley, USA

A7 Adhesive layer ultimate strength approach using fracture mechanics

The results of the elastic stress analysis approach described in Appendix 5 are sensitive to the thickness and the shear modulus of the adhesive. Due to the variability in these parameters the peak shear stress is subject to considerable variability. Its percentage variability is 50 per cent of that of the adhesive thickness. For example, if the adhesive thickness varies between 0.5 mm and 1.0 mm, the shear stress might vary by 25–50 per cent. The ultimate shear strength of the adhesive is also variable, being dependent upon the extent of the stress concentration zone and the presence of flaws in the adhesive. The capacity of the joint is a function of the margin between adhesive strength and the peak shear stress and is therefore subject to a greater variability than either of these variables. As a result, there is considerable inherent uncertainty in a joint capacity prediction based on the elastic stress-based analysis. This necessitates a sufficiently high partial safety factor.

An alternative approach is to assess the ultimate capacity of the joint on the basis of ultimate energy that the adhesive joint can absorb during failure by cracking. The ultimate energy release rate G associated with the propagation of a crack in the adhesive layer is used to characterise the strength of the joint.

Since this method eliminates from consideration the elastic stress distribution in the adhesive layer, the equations are simpler to use than those of the elastic stress based approach.

The fracture energy release rate is defined by

$$G_I = -\frac{1}{b_a}\frac{\partial U}{\partial a} \tag{A7.1}$$

where b_a is the width of the adhesive layer, U is the total strain energy of the beam, and a is the length of a delamination crack between the beam and the strengthening plate. For a beam and strengthening plate of constant section subject to a uniform load the net strain energy is:

$$U = u_2 a + u_1(L - a) \tag{A7.2}$$

u_1 and u_2 are the strain energy per unit length of the strengthened and unstrengthened beam sections respectively (see Figure A7.1). It follows that the energy release rate G_I is:

$$G_I = \frac{1}{b_a}(u_1 - u_2) \tag{A7.3}$$

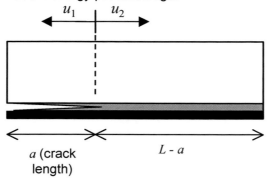

Strain energy per unit length

Figure A7.1 *Strain energy in a strengthened section as a crack propagates through the adhesive layer*

For the crack to be stable:

$$G_I \leq \overline{G}_{IC} \tag{A7.4}$$

$\overline{G}_{IC}$ is the critical energy release rate which characterises the strength, or more appropriately toughness, of the adhesive joint. $\overline{G}_{IC}$ can be determined using standard fracture mechanics tests, such as the double cantilever beam (DCB) test (BS 7991:2001)

Mode I failure (peeling) is assumed to dominate failure of the adhesive joint. This will usually be the case for an adhesive joint, as the critical energy release rate for Mode II failure (shearing) is much higher than that for mode I ($\overline{G}_{IC} \ll \overline{G}_{IIC}$).

The strain energy per unit length in a beam subjected to an axial force N at the tension face of a section and a bending moment M is given by:

$$u = \frac{1}{2}\frac{N^2}{(EA)_S} + \frac{1}{2}\frac{(M - N y_{g0})^2}{(EI)_S} \tag{A7.5}$$

where $(EA)_S$, $(EI)_S$ are the section stiffness properties of the beam section with respect to the centre of stiffness of the beam, at level y_{g0}. These are defined by:

$$(EA)_S = \int_A E_S \, dA \tag{A7.6}$$

$$y_{g0} = \frac{1}{(EA)_S} \int_A E_S y \, dA \tag{A7.7}$$

$$(EI)_S = \int_A E_S (y - y_{g0})^2 \, dA \tag{A7.8}$$

For a beam strengthened with FRP material, the properties of the composite section are:

$$(EA)_1 = E_S A_S + E_f A_f \tag{A7.9}$$

$$y_{g1} = \frac{1}{(EA)_1}(E_S A_S y_S + E_f A_f y_f) \tag{A7.10}$$

$$(EI)_1 = E_S \left(I_S + A_S \left(y_S - y_{g1} \right)^2 \right) + E_f A_f \left(y_f - y_{g1} \right)^2 \qquad \text{(A7.11)}$$

E_S, A_S, y_S, I_S are the modulus, sectional area, centroidal position, and second moment of area of the unstrengthened beam, and E_f, A_f, y_f are the modulus, cross-sectional area and centroidal position of the FRP strengthening material.

For the simple case of a beam subject to a moment of M:

$$u_1 = \frac{M^2}{2(EI)_S} \qquad \text{(A7.12)}$$

$$u_2 = \frac{M^2}{2(EI)_1} \qquad \text{(A7.13)}$$

$(EI)_1$ and $(EI)_s$ are the flexural stiffness properties of the composite, strengthened section and the unstrengthened section respectively. Therefore

$$G = \frac{M^2}{2b_a} \left(\frac{1}{(EI)_S} - \frac{1}{(EI)_1} \right) \qquad \text{(A7.14)}$$

A7.1   **TEMPERATURE EFFECTS**

Consider a beam subjected to a through-thickness temperature distribution $T(y)$, as shown in Figure A7.2. The fully released strain distribution is:

$$\varepsilon = \varepsilon_1 + \psi_1 (y - y_g) - \alpha T \qquad \text{(A7.15)}$$

where ε_1 is the strain at the level of the centroid, ψ_1 is the curvature, α is the coefficient of thermal expansion, y is the vertical position, y_{g1} is the position of the centroid, and T is the operating temperature.

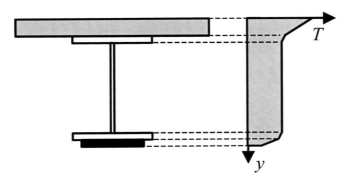

Figure A7.2 *Schematic temperature distribution through the strengthened section*

The fully restrained thermal resultants are:

$$N_t = \int_A E \alpha T \, dA \qquad \text{(A7.16)}$$

$$M_t = \int_A E \alpha T \, y \, dA \qquad \text{(A7.17)}$$

For the case of no external applied loads (ie $N = 0$, $M = 0$), the fully released strain parameters are:

$$\varepsilon_1 = \frac{N_t}{(EA)_1} \tag{A7.18}$$

$$\psi_1 = \frac{M_t - N_t y_{g1}}{(EI)_1} \tag{A7.19}$$

The strain energy of the section per unit length is given by:

$$u = \frac{1}{2} \int_A E \left(\varepsilon_1 + \psi_1 (y - y_{g1}) - \alpha T \right)^2 dA \tag{A7.20}$$

which equates to:

$$u = \frac{1}{2} \left(L_0 - N_t \varepsilon_1 - M_t \psi_1 \right) \tag{A7.21}$$

where:

$$L_0 = \int_A E (\alpha T)^2 dA \tag{A7.22}$$

Since the fully released thermal stresses in an unstrengthened beam are normally *lower* than those in a beam strengthened with a material of different coefficient of thermal expansion, the strain energy of the unstrengthened beam is also lower. Therefore, the fracture energy release rate for the temperature case is negative. This indicates that, in the case of an imposed strain load case, cracking of the bond between the strengthening plate and the beam results in relaxation of the stresses.

A7.2   COMBINED EXTERNAL LOADING AND TEMPERATURE

The strain components for combined external loading and temperature are

$$\varepsilon_1 = \frac{N + N_t}{(EA)_1} \tag{A7.23}$$

$$\psi_1 = \frac{M + M_t - (N + N_t) y_{g1}}{(EI)_1} \tag{A7.24}$$

The strain energy per unit length is

$$u = \frac{1}{2} \left\{ L_0 + (N - N_t) \varepsilon_1 + (M - M_t) \psi_1 \right\} \tag{A7.25}$$

A8 Experimental determination of the variability partial factor

There is a wide range of composite materials available, each with a number of independent material properties. To characterise the variability partial factor for a particular material, it is first necessary to determine the statistical distribution of the material property of interest. This requires test results from a statistically significant number of tests on a representative sample of the material.

From the test results the mean and standard deviation are determined. The reference value adopted for the material property is the characteristic value, conventionally defined as the value that has a 95 per cent probability of being exceeded. A normal distribution is assumed, so that the characteristic value, is given by:

$$X_k = \mu_X - 1.65\sigma_X \qquad (A8.1)$$

where μ_X and σ_X are the mean and standard deviation of property X.

The variability partial factor, γ_{mv}, may be determined from the statistical parameters using:

$$\gamma_{X_i} = \frac{1 + \xi_{X_i u} V_{X_i}}{1 + \xi_{X_i k} V_{X_i}} \qquad (A8.2)$$

where:

$$\xi_{X_i u} = -\alpha_i \beta \qquad (A8.3)$$

γ_{Xi} is the partial safety factor (if $\gamma_{Xi} < 1$ as in the case of resistance variables the reciprocal is specified in the codes)

V_{Xi} is the coefficient of variation defined by:

$$V_X = \frac{\sigma_X}{\mu_X} \qquad (A8.4)$$

$\xi_{X,k}$ is the non-dimensional variate corresponding to the characteristic value (normally -1.65, as in Equation A8.1)

$\xi_{X,u}$ is the non-dimensional variate corresponding to the ultimate value

β is the target reliability

α_i is the direction cosine of the unit normal to the limit state function given by:

$$\alpha_i = \frac{\partial g}{\partial x_i} \Big/ \sqrt{\sum_j \left(\frac{\partial g}{\partial x_j}\right)^2} \qquad (A8.5)$$

For a linear limit state function $\alpha \cong 0.75$. As a first approximation the $\xi_{X_i u}$ value can be derived from another material for which V_{Xi} and γ_{Xi} are known:

$$\xi_{X_i u} = \frac{\gamma_{X_i}\left(1 + \xi_{X_i k}V_{X_i}\right) - 1}{V_{X_i}}$$

(A8.6)

and if necessary modified to allow for a different value of β.

The relationship between the partial factor (γ) and coefficient of variation (V_{Xi}) shown in Figure A8.1 was obtained using this method, assuming an extremal type I distribution for X_i, and with concrete used as a reference to provide the value of β.

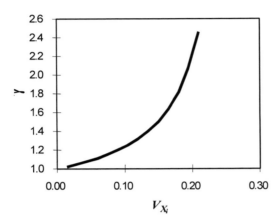

Figure A8.1 *Indicative partial resistance factor versus COV*

A8.1   BIBLIOGRAPHY

Cadei, J M C (1998). "Factors of safety in the limit state design of FRP composite structures". In: *Proc designing cost-effective composites, Sep, IMechE, London*. Professional Engineering Publishing, London, pp 27–40

CIRIA (1977). *Rationalisation of safety and serviceability factors in structural codes*. Report 63, CIRIA, London

A9   Worked example

The worked example presented in this appendix aims to provide the reader with confidence in using the analysis methods given in Chapter 5. It covers the core analysis required for the FRP strengthening for metallic structures:

- sectional design (Section 5.2)
- adhesive joint design (Section 5.3).

The wider design process is not considered; for example, no partial factors are applied, and a specified FRP arrangement is analysed.

The example should be read in conjunction with Chapter 5. It is laid out in the same order, and references are made to the sections and equation numbers in that chapter.

A9.1   DEFINITION OF EXAMPLE

Figure A9.1 shows the geometry of a simply supported cast-iron beam, which requires strengthening to increase its moment capacity. Preformed FRP plates with unidirectional carbon fibres will be prestressed and bonded to the soffit of the beam. Table A9.1 gives the properties of the materials, and Table A9.2 gives geometric parameters that have been derived from the basic geometry.

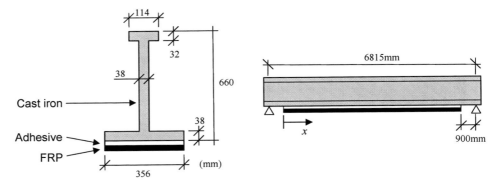

Figure A9.1 *Geometry of cast-iron beam considered in the worked example*

Table A9.1 *Material properties for worked example*

Cast iron

Young's modulus	E_s	138 kN/mm^2
Coefficient of thermal expansion	α_s	11×10^{-6} /°C
Permissible tensile strains	$\overline{\varepsilon}_{st}$	Permanent stress = 19.0 N/mm^2, 0.0255%
		Permanent stress = 14.4 N/mm^2, 0.0120%

CFRP strengthening

Young's modulus	E_f	360 kN/mm^2
Coefficient of thermal expansion	α_f	1×10^{-6} /°C
Breadth	b_f	356 mm
Thickness	t_f	11 mm
Prestress (before transfer)	σ_{fp}	150 MPa
Permissible tensile strain	$\overline{\varepsilon}_{ft}$	0.1 per cent

---

Adhesive

Young's modulus	E_a	4.5 kN/mm^2
Shear modulus	G_a	1.7 kN/mm^2
Thickness	t_a	2 mm
Breadth	b_a	356 mm

Note: The permissible tensile strain in the cast iron is based on BD 21/01.

Table A9.2 *Derived section properties for worked example*

Cast iron section

Height of centroid	y_{g0}	254 mm
Distance from centroid to adhesive interface ($= y_{g0}$ for this case)	y_s	254 mm
Cross-sectional area	A_s	39 596 mm^2
Second moment of area	I_s	2.094×10^9 mm^4

CFRP strengthening

Distance from centroid to adhesive interface	y_f	5.5 mm
Cross-sectional area	A_f	3916 mm^2
Second moment of area	I_f	39 486 mm^4
Internal lever arm	$z = y_f + y_s + t_a$	261.5 mm

The span of the beam is subjected to uniformly distributed load:

- dead load (present at the time of strengthening), $w_D = 27$ kN/m
- live load (applied after strengthening), $w_L = 35.1$ kN/m

The temperatures of the cast iron and FRP are assumed to be uniform:

- temperature at time of installation of cast iron beam, $T_i = 20$°C
- temperature at time of strengthening, $T_0 = 20$°C
- maximum operating temperature, $T = 50$°C
- minimum operating temperature, $T = -10$°C

The bending moment distribution along the beam due to uniformly distributed load is:

$$M = \frac{w}{2}\left[(x+900)L - (x+900)^2\right] \quad \text{(kNmm)} \quad \text{(sagging } +ve\text{)} \tag{A9.1}$$

Note that $x = 0$ is defined as the end of the strengthening plate (Figure A9.1), as required in the adhesive joint analysis.

A9.2   SECTIONAL ANALYSIS (SECTION 5.2.3)

The FRP required is sized at the location of the maximum moment, by considering a section through the member at mid-span. At mid-span:

     Dead load moment $M_D = 156.7$ kNm

     Live load moment $M_L = 203.8$ kNm

     Dead + live moment $M_U = 360.5$ kNm

Step 1 – Determine the capacity of the unstrengthened member

(5.11) $M_{su} = \dfrac{(EI)_s \bar{\varepsilon}_{st}}{y_{g0}}$ $\boxed{M_{su} = 289.7 \text{ kNm}}$

Note that this is less than the dead + live load moment that the beam is required to carry.

Step 2 – Determine the state of stress and strain in the unstrengthened member at the time of application of the strengthening

In this example there is no change in temperature between installation of the beam and the application of strengthening, ie $T_0 - T_i = 0$ in (5.14).

(5.14) $N_{t0} = -(EA)_s \alpha_s (T_0 - T_i)$ $\boxed{N_{t0} = 0 \text{ kN}}$

(5.15) $M_{t0} = 0$ $\boxed{M_{t0} = 0 \text{ kNm}}$

(N_0, M_0) include:

- the stress resultants due to permanent load, present at the time of strengthening
- the stress resultants in the metallic member due to prestress in the FRP
- the stress resultants due to statically indeterminate effects of thermal loading.

The dead load acts on the beam at the time of strengthening, giving a mid-span moment of 156.7×10^3 kNmm.

For this example, it is assumed that the prestress is applied to the FRP using jacks that are anchored to the cast iron member. For a prestress of $\sigma_{fp} = 150$ MPa, the axial load in the cast iron is $-A_f.\sigma_{fp} = -587.4$ kN (compression), and the moment is $-A_f.\sigma_{fp}.(y_{g0}+0.5t_f) = -152.7 \times 10^3$ kNmm (hogging).

The net stress resultants in the metallic member are thus:

- $N_0 = -587.4$kN
- $M_0 = 156.7 \times 10^3 - 152.7 \times 10^3 = 4 \times 10^3$ kNmm

$$（5.16） \quad \varepsilon_{s0} = \frac{N_0 - N_{t0}}{(EA)_s} + \frac{M_0 - M_{t0}}{(EI)_s} y_{g0}$$

$$\boxed{\varepsilon_{s0} = -0.0104\ \%}$$

$$（5.17） \quad \sigma_{s0} = E_s \left\{ \frac{N_0 - N_{t0}}{(EA)_s} + \frac{M_0 - M_{t0}}{(EI)_s} y_{g0} - \alpha_s (T_0 - T_i) \right\}$$

$$\boxed{\sigma_{s0} = -14.4\ \text{N/mm}^2}$$

Step 3 – Establish limiting strains in the strengthened member

The prestrain in the FRP, $\varepsilon_{fp} = \sigma_{fp} / E_f = 0.0417\%$

$$（5.19） \quad \varepsilon_{f0} = \varepsilon_{s0} - \varepsilon_{fp}$$

$$\boxed{\varepsilon_{f0} = -0.0521\ \%}$$

The limiting strains adjacent to the adhesive interface are:

- $\bar{\varepsilon}_{st} = 0.0120\%$ in the substrate (for a stress under permanent loads of -14.4 N/mm^2)

- $\bar{\varepsilon}_{ft} = 0.1\%$ in the FRP, giving $\bar{\varepsilon}_{ft} + \varepsilon_{f0} = 0.0479\%$

$$（5.18） \quad \bar{\varepsilon}_{yg0} = \min(\bar{\varepsilon}_{st},\ \bar{\varepsilon}_{ft} + \varepsilon_{f0})$$

$$\boxed{\bar{\varepsilon}_{yg0} = 0.0120\%}$$

Step 4 – Initial sizing of the FRP strengthening material

In this example, the thickness of FRP has been specified. However, the amount of non-prestressed FRP can be estimated, by using Equation 5.20 with $M_0 = 156.7 \times 10^3$ kNmm (for no prestress):

$$（5.20） \quad A_f = \frac{E_s}{E_f} \left(\frac{M_u - M_{su}}{M_{su} - M_0} \right) \left(\frac{1}{1 + (y_{g0}^2 A_s / I_s)} \right) A_s$$

$$\boxed{A_f = 3638\ \text{mm}^2}$$
(for unstressed strengthening)

A 10.2 mm-thick strengthening plate would thus be required if differential thermal effects are neglected and the FRP was not prestressed. Prestressed strengthening (or a load-relief jacking method) is likely to be more economic, particularly when differential thermal effects are considered.

Step 5 – Section properties of the strengthened member

$$（5.21） \quad (EA)_1 = (EA)_s + (EA)_f$$

$$\boxed{(EA)_1 = 6.869 \times 10^6\ \text{kN}}$$

$$（5.22） \quad y_{g1} = \frac{y_{g0}(EA)_s - 0.5 t_f (EA)_f}{(EA)_1}$$

$$\boxed{y_{g1} = 201\ \text{mm}}$$

$$（5.23） \quad (EI)_1 = (EI)_s + (EA)_s (y_{g0} - y_{g1})^2 + (EA)_f (y_{g1} + 0.5 t_f)^2$$

$$\boxed{(EI)_1 = 364.7 \times 10^9\ \text{kNmm}^2}$$

Step 6 – Calculate strains in the strengthened member

The strengthened member is subjected to a uniform temperature decrease, $(T_s - T_i) = -30°C$ in both FRP and the metallic member.

The total applied loads (dead AND live) acting on the beam at mid-span are $N = 0$ and $M = 360\ 500$ kNmm

(5.26) $\quad N_t = -(EA)_s \alpha_s (T_s - T_i) - (EA)_f \alpha_f (T_f - T_{f0})$ $\qquad \boxed{N_t = 1844\ \text{kN}}$

(5.27) $\quad M_t = -(EA)_f \alpha_f (T_f - T_{f0})(y_{g0} + 0.5t_f)$ $\qquad \boxed{M_t = 10992\ \text{kNmm}}$

(5.28) $\quad \varepsilon_1 = \dfrac{N - N_t + (EA)_f \varepsilon_{f0}}{(EA)_1}$ $\qquad \boxed{\varepsilon_1 = -0.0375\ \%}$

(5.29) $\quad \psi_1 = \dfrac{(M - M_t) - (N - N_t)(y_{g0} - y_{g1}) + (EA)_f \varepsilon_{f0}(y_{g1} + 0.5t_f)}{(EI)_1}$ $\qquad \boxed{\psi_1 = 0.81 \times 10^{-6}\ \text{mm}^{-1}}$

Step 7 – Check strains against limiting strains

Strain in substrate:

(5.30) $\quad \varepsilon_1 + \psi_1 y_{g1} - \alpha_s (T_s - T_i) \le \overline{\varepsilon}_{st}$ $\qquad \boxed{0.0118\% < 0.0120\%}$
(OK)

Strain in FRP:

(5.31) $\quad \varepsilon_1 + \psi_1 y_{g1} - \alpha_f (T_f - T_{f0}) - \varepsilon_{f0} \le \overline{\varepsilon}_{ft}$ $\qquad \boxed{0.0339\ \% < 0.1\ \%}$
(OK)

The specified quantity of strengthening is thus suitable for the applied loads and operating temperature.

The strains calculated above are due to the combined effects of:

- prestrains due to prestressing the FRP, and the permanent loads acting at the time of strengthening
- applied live load
- differential thermal expansion.

Table A9.3 compares these different components of strain, showing that the temperature strains are of similar magnitude to the strains due to the applied live load. (These strains were calculated by repeating the above calculations with the appropriate live loads.)

Table A9.3   *Strains adjacent to adhesive interface due to applied loads*

	Dead load only	Prestress	Live load	± temperature differential	Max strain
Strain in substrate	0.0086%	-0.0241%	0.0112%	±0.0110%	0.0118 %
Strain in FRP	-0.0001%	0.0418%	0.0112%	±0.0190 %	0.0719 %

A9.3   ADHESIVE JOINT ANALYSIS (SECTION 5.3.1)

An elastic, stress-based analysis is used here to assess the stresses in the adhesive joint.

Step 1 – Determine strength of the adhesive from the lap-shear test result

A set of single sided lap-shear tests was used to determine the strength of the adhesive joint. The specimens were fabricated according to BS 5350-C5:2002 (see Section 7.1.1), using the CFRP and adhesive used to strengthen the structure. The parameters not defined in Table A9.1 are:

Length of overlap	$L_a = 25$ mm	
Width of joint	$b_a = 25$ mm	
Thickness of FRP	$t_f = 3$ mm	
Poisson's Ratio of FRP	$v_f = 0.3$	

The characteristic failure load was 18.25 kN (an *average* shear stress of 29.2 N/mm^2 along the lap-shear specimen).

(A6.3) $\quad c_a = \dfrac{L_a}{2}$ $\qquad\qquad$ $\boxed{c_\mathrm{a} = 12.5 \text{ mm}}$

(A6.4) $\quad \beta_s = \sqrt{8\dfrac{G_\mathrm{a}\, t_f}{E_f\, t_\mathrm{a}}}$ $\qquad\qquad$ $\boxed{\beta_\mathrm{s} = 0.238}$

(A6.5) $\quad \lambda_s = \dfrac{c_a}{t_f}\left(\dfrac{6E_\mathrm{a}t_f}{E_f t_\mathrm{a}}\right)^{\frac{1}{4}}$ $\qquad\qquad$ $\boxed{\lambda_\mathrm{s} = 2.41}$

(A6.9) $\quad u_4 = \dfrac{1}{\sqrt{2}t_f}\sqrt{3\left(1 - v_f^{\,2}\right)\dfrac{P}{b_\mathrm{a}t_f E_f}}$ $\qquad\qquad$ $\boxed{u_4 = 0.0101}$

(A6.8) $\quad u_3 = 2\sqrt{2}u_4$ $\qquad\qquad$ $\boxed{u_3 = 0.0286}$

(A6.6) $\quad k = \dfrac{\cosh(u_4 c_a)\sinh(u_3 L_a)}{\sinh(u_3 L_a)\cosh(u_4 c_a) + 2\sqrt{2}\cosh(u_3 L_a)\sinh(u_4 c_a)}$ $\qquad$ $\boxed{k = 0.633}$

(A6.7) $\quad k' = k\dfrac{c_a}{t_f}\left(3\left(1 - v_f^{\,2}\right)\dfrac{P}{b_\mathrm{a}t_f E_f}\right)^{\frac{1}{2}}$ $\qquad\qquad$ $\boxed{k' = 0.113}$

The peak shear and peel stress in the lap-shear specimen are:

$$(A6.1) \quad \tau_{max} = \frac{P}{8b_a c_a}\left[\begin{array}{l}\dfrac{\beta_s c_a}{t_f}(1+3k)\coth\left(\dfrac{\beta_s c_a}{t_f}\right)\\ +3(1-k)\end{array}\right] \quad \left(\text{for } \frac{\beta_s c_a}{t_f}<25\right) \qquad \boxed{\tau_{max}=35.7 \text{ N/mm}^2}$$

$$(A6.2) \quad \sigma_{max} = \frac{P}{b_a t_f}\left(\frac{t_f}{c_a}\right)^2\left[\begin{array}{l}\lambda_s^2\,\dfrac{k}{2}\,\dfrac{\sinh(2\lambda_s)-\sin(2\lambda_s)}{\sinh(2\lambda_s)+\sin(2\lambda_s)}\\ -\lambda_s k'\dfrac{\cosh(2\lambda_s)+\cos(2\lambda_s)}{\sinh(2\lambda_s)+\sin(2\lambda_s)}\end{array}\right] \quad (\text{for } \lambda_s<2.5) \qquad \boxed{\sigma_{max}=22.8 \text{ N/mm}^2}$$

The characteristic strength of the adhesive is:

$$(A6.10) \quad \bar{\sigma} = \frac{\sigma_{max}}{2}+\sqrt{\left(\frac{\sigma_{max}}{2}\right)^2+\tau_{max}^2} \qquad\qquad \boxed{\bar{\sigma}=48.9 \text{ N/mm}^2}$$

Step 2 – Calculate the maximum adhesive shear stress

Lack-of-fit strain, $\Delta\varepsilon_{fs}$

The peel stress analysis (below) requires the lack-of-fit strain to be differentiated with respect to position along the adhesive joint, thus the variation of $\Delta\varepsilon_{fs}$ with x is required. In the present case, only the moment applied after strengthening, M_s, varies along the beam:

$$(A9.1a) \quad M_s(x) = -\frac{w_L}{2}\left[(x+900)L-(x+900)^2\right] \quad \text{(hogging +ve)}$$

Axial load applied after strengthening	$N_s = 0$ kN
Axial load due to prestress	$N_{f0} = 150 \text{ N/mm}^2 \times 3916 \text{ mm}^2 = 587.4\text{kN}$
Moment due to pre-curvature	$M_{f0} = 0$ kNmm
Temperature change after strengthening (for both substrate and FRP)	$(T-T_0) = 50-20 = 30°C$

$$(5.40) \quad \psi_{sR} = \frac{M_s}{E_s I_s}$$

$$(5.39) \quad \varepsilon_{sR} = \frac{N_s}{E_s A_s}-\psi_{sR}y_s+\alpha_s\left(T_s-T_{s0}\right)$$

$$(5.42) \quad \psi_{fR} = -\frac{M_{f0}}{E_f I_f}$$

$$(5.41) \qquad \varepsilon_{fR} = -\frac{N_{f0}}{E_f A_f} + \psi_{fR} y_f + \alpha_f (T_f - T_{f0})$$

$$(5.43) \qquad \Delta\varepsilon_{fs}(x) = \varepsilon_{fR} - \varepsilon_{sR}$$

Substituting into Equation 5.39 to 5.43 gives:

$$
\begin{aligned}
\Delta\varepsilon_{fs}(x) &= \frac{y_s}{E_s I_s} M_s(x) - \frac{N_{f0}}{E_f A_f} + (\alpha_f - \alpha_s)(T - T_0) \\
&= [0.8803 M_s(x) - 709.3 N_{f0} - 10000(T - T_0)] \times 10^{-9}
\end{aligned}
\qquad (A9.2)
$$

Flexibility parameters

$$(5.44) \qquad f_1 = \left(\frac{t_a}{G_a b_a}\right) \qquad\qquad \boxed{f_1 = 0.0033 \text{ mm}^2/\text{kN}}$$

$$(5.45) \qquad f_2 = \left(\frac{1}{E_f A_f} + \frac{1}{E_s A_s} + \frac{z(z - t_a)}{E_s I_s + E_f I_f}\right) \qquad \boxed{f_2 = 1.128 \times 10^{-6} \text{ kN}^{-1}}$$

$$(5.46) \qquad \lambda = \sqrt{\frac{f_2}{f_1}} \qquad\qquad \boxed{\lambda = 0.0185 \text{ mm}^{-1}}$$

Maximum adhesive shear stress

The axial force is zero at the end of the FRP, hence $N_f\big|_{x=0} = 0$

The lack-of-fit strain at the end of strengthening ($x = 0$) is $\Delta\varepsilon_{fs}\big|_{x=0} = -0.0799\,\%$

The peak shear stress due to geometric discontinuity at the end of the FRP is:

$$(5.47) \qquad \tau_{\max} = -\frac{\Delta\varepsilon_{fs}\big|_{x=0}}{b_a \sqrt{f_1 f_2}} - \frac{\lambda N_f\big|_{x=0}}{b_a} \qquad \boxed{\tau_{\max} = 36.8 \text{ N/mm}^2}$$

The shear force acting on the strengthened member, $V = dM_s/dx = -88$ kN (by differentiating Equation A9.1 at x = 0). The flexural shear stress is:

$$(5.48) \qquad \tau = \frac{V}{b_a} \frac{E_f A_f y_{g1}}{(EI)_1} \qquad\qquad \boxed{\tau = 0.19 \text{ N/mm}^2}$$

The total peak shear stress in the adhesive joint is the sum of Equations 5.47 and 5.48, giving 37.0 N/mm². As noted in Section 5.3.1, the shear stresses due to flexure of the beam are small compared with the concentration of shear stress at the end of the FRP.

Step 3 – Calculate the maximum adhesive peel stress

Lack-of-fit curvature, $\Delta \psi_{\text{fs}}$

The variation in lack-of-fit curvature is required with x (as for the lack-of-fit strain, above):

(5.49) $\qquad \Delta \psi_{\text{fs}}(x) = \psi_{\text{fR}} - \psi_{\text{sR}}$

$$\begin{aligned} \Delta \psi_{\text{fs}}(x) &= -\frac{M_s}{E_s I_s} \\ &= -3.46 \times 10^{-12} M_s(x) \end{aligned}$$

(A9.3)

Flexibility parameters

(5.50) $\qquad a_1 = \dfrac{t_{\text{a}}}{E_{\text{a}} b_{\text{a}}}$ $\qquad\qquad\qquad\qquad$ $\boxed{a_1 = 1.248 \times 10^{-3} \ \text{mm}^2/\text{kN}}$

(5.51) $\qquad a_2 = \dfrac{1}{E_f I_f} + \dfrac{1}{E_s I_s}$ $\qquad\qquad\qquad$ $\boxed{a_2 = 70.4 \times 10^{-9} \ \text{mm}^{-2} \text{kN}^{-1}}$

(5.52) $\qquad a_3 = \dfrac{z - y_f}{E_s I_s} - \dfrac{y_f}{E_f I_f}$ $\qquad\qquad\quad$ $\boxed{a_3 = -386 \times 10^{-9} \ \text{mm}^{-1} \text{kN}^{-1}}$

(5.53) $\qquad \beta = \left(\dfrac{a_2}{4 a_1} \right)^{0.25}$ $\qquad\qquad\qquad\quad$ $\boxed{\beta = 0.0613 \ \text{mm}^{-1}}$

Moment due to lack of fit and the axial force in the strengthening

(5.55) $\qquad C_4 = -\dfrac{1}{a_2} \left(\Delta \psi_{\text{fs}}(x) + a_3 \dfrac{\Delta \varepsilon_{\text{fs}}(x)}{f_2} \right)$

(5.56) $\qquad C_5 = \dfrac{a_3}{a_1 \lambda^4 + a_2} \left[N_f \big|_{x=0} + \dfrac{\Delta \varepsilon_{\text{fs}} \big|_{x=0}}{f_2} \right]$ $\qquad$ $\boxed{C_5 = 3878 \ \text{kNmm}}$

(5.54) $\qquad \begin{aligned} M_{\text{PS}}^* &= C_4(x) + C_5 e^{-\lambda x} \\ &= 4.33 \times 10^{-3} M_s - 3.45 N_{f0} - 48.6(T - T_0) + 3878 e^{-\lambda x} \end{aligned}$

Maximum adhesive peel stress

(5.58) $\qquad C_2 = -M_{\text{PS}}^* \big|_{x=0}$ $\qquad\qquad\qquad\qquad$ $\boxed{C_2 = 12.6 \ \text{kNmm}}$

Substituting Equation A9.1 into the expression for $M_{PS}{}^*$, and differentiating gives:

$$\frac{\mathrm{d}M_{PS}^*}{\mathrm{d}x} = -4.33 \times 10^{-3} \frac{w_L}{2}[L - 2(x+900)] - 3878\lambda e^{-\lambda x} \qquad \boxed{\left.\frac{\mathrm{d}M_{PS}^*}{\mathrm{d}x}\right|_{x=0} = -72.03 \text{ kN}}$$

(5.59) $\quad C_3 = -\dfrac{1}{\beta}\left[\left.\dfrac{\mathrm{d}M_{PS}^*}{\mathrm{d}x}\right|_{x=0} - \beta C_2\right]$ $\qquad \boxed{C_3 = 1188 \text{ kNmm}}$

Differentiating $\mathrm{d}M_{PS}{}^*/\mathrm{d}x$ to find $\mathrm{d}^2 M_{PS}{}^*/\mathrm{d}x^2$ gives:

$$\frac{\mathrm{d}^2 M_{PS}^*}{\mathrm{d}x^2} = \frac{\mathrm{d}^2 C_4}{\mathrm{d}x^2} \qquad + \lambda^2 C_5 e^{-\lambda x}$$
$$= 4.33 \times 10^{-3} w_L + 3878\lambda^2 \qquad \text{at } x = 0$$
$$= 0.00015 \qquad\quad + 1.33$$

As discussed in Appendix 5 (above Equation A5.37), $\mathrm{d}^2 C_4/\mathrm{d}x^2$ is negligible compared with $\lambda^2 C_5$, and the simplified equation for the peak peel stress can be used. The maximum peel stress in the adhesive is thus:

(5.57) $\quad \sigma_{max} = -\dfrac{1}{b_a}\left[C_5\lambda^2 - 2C_3\beta^2\right]$ $\qquad \boxed{\sigma_{max} = 21.3 \text{ N/mm}^2}$

Step 4 – Calculate the principal stress and compare with the strength of the adhesive

The principal stress is found from the combination of the peak shear stress and the peak peel stress in the adhesive joint.

(5.38) $\quad \sigma_1 = \dfrac{\sigma_{max}}{2} + \sqrt{\left(\dfrac{\sigma_{max}}{2}\right)^2 + \tau_{max}{}^2} \leq \bar{\sigma}$ $\qquad \boxed{\sigma_1 = 49.2 \text{N/mm}^2}$

From Step 1, the characteristic strength of the adhesive joint is $\bar{\sigma} = 48.9\,\text{N/mm}^2$, which is just less than the peak principal stress that the joint is required to carry. The high peak stress is due in part to the prestress in the strengthening. A mechanical anchorage is usually used to clamp the end of the strengthening if it is prestressed, especially since large permanent stresses are carried through the adhesive.

Stress distributions in the adhesive

Figure A9.2 shows the distributions of shear and peel stress along the adhesive joint for different load cases. As in the sectional analysis, the adhesive stresses due to temperature changes are large compared with those due to the action of the live load.

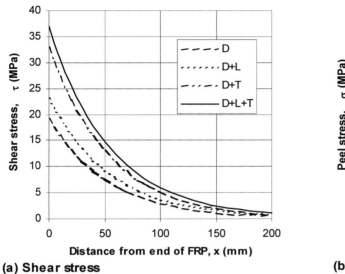

(a) Shear stress

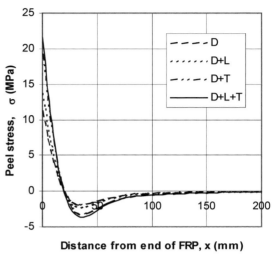

(b) Peel stress

D = Dead load (present at time of strengthening), including prestress

L = Live load

T = Temperature

Figure A9.2 *Shear stress and peel stress distributions in the adhesive joint for different load cases*

Consolidated bibliography

ACI (2002). *Guide for the design and construction of externally bonded FRP systems for strengthening concrete structures.* ACI 440.2R-02, American Concrete Institute, Detroit

Adams, R D, Comyn, A and Wake, W C (1997). *Structural adhesive joints in engineering.* 2nd edn, Chapman and Hall

Albat, A M and Romilly, D P (1999). "A direct linear-elastic analysis of double symmetric bonded joints and reinforcements". *Composite science and technology,* vol 59, no 7, pp 1127–1137

Angus, H T (1976). *Cast iron: Physical and engineering properties.* 2nd edn, Butterworth, London

ASCE (1984). *Structural plastics design manual.* ASCE Manuals & Reports on Engineering Practice No 63, American Society of Civil Engineers, Reston, VA

Aylor, D M (1993). "The effect of a seawater environment on the galvanic corrosion behavior of graphite/epoxy composite coupled to metals". In: C E Harris and T S Gates (eds), *High temperature and environmental effects on polymeric composites.* STP 1174, ASTM, West Conshohocken, PA

Baker, A (1996). "Joining and repair of aircraft composite structures". *Mech engg trans,* vol ME21, nos 1 and 2

<http://www.balvac.co.uk/experience/projects/project_h4.html>

<http://www.balvac.co.uk/experience/projects/project_s2.html>

Bank, L C and Mosallam, A S (1992). "Creep and fatigue of a full-size fibre-reinforced plastic pultruded frame". *Comp engg,* vol 2, no 3, pp 213–227

Barnes, R A and Mays, G C (2001). "The effect of traffic vibration on adhesive curing during installation of bonded external reinforcement". *Proc Inst Civ Engrs, Structures and buildings,* vol 136, no 4, pp 403–410

Barnfield, J R and Porter, A M (1984). "Historic buildings and fire: fire performance of cast iron structural element". *The structural engineer,* vol 62A, no 12, pp 373–380

Bassetti, A (2001). "Lamelles précontraintes en fibres carbone pour le renforcement de ponts rivetés endommagés par fatigue". PhD thesis EPFL 2440, École Polytechnique Fédérale, Lausanne

Bassetti, A, Nussbaumer, A and Colombi, P (2000). "Repair of riveted bridge members damaged by fatigue using CFRP materials". In: G Pascale (ed), *Proc conf advanced FRP materials for civil structures, Bologna,* pp 33–42

Bassetti, A, Nussbaumer, A and Hirt, M (2000). "Crack repair and fatigue life extension of riveted bridge members using composite materials". In: A-H Hosny (ed), *Proc bridge engg conf 2000, ESE-IABSE-FIB, Sharm El-Sheikh*. Egyptian Soc Engrs, Cairo, vol I, pp 227–238

Bates, W (1984). *Historical structural steelwork handbook*. BCSA 11/84, British Constructional Steel Association, London

Bellucci, F (1992). "Galvanic corrosion between nonmetallic composites and metals II. Effect of area ratio and environmental degradation". *Corrosion*, vol 48, pp 281–291

Blontrock, H, Taerwe, L and Vandevelde, P (2001). "Fire testing of concrete slabs strengthened with fibre composite laminates". In: *Proc 5th int conf FRP reinf conc strucs (FRPRCS-5), 16–18 Jul, Cambridge*. Thomas Telford, London, pp 547–556

Bowditch, M R (1996). "The durability of adhesive joints in the presence of water". *Int j adhesion and adhesives*, vol 16, no 2, pp 73–79

Boyd, J et al (1991). "Galvanic corrosion effects on carbon fiber composites". *Proc 36th int SAMPE symp, 14–18 Apr, San Diego*. SAMPE, Covina, CA

Brinson, H P, Morris, D H and Yeow, Y T (1978). "A new experiment method for accelerating characterisation of composite materials". *Proc 6th int conf experimental stress analysis, 18–22 Sep, Munich*, p 395

Bussell, M N (1997). *Appraisal of existing iron and steel structures*. P138, Steel Construction Institute, Ascot

Bussell, M N and Robinson, M J (1998). "Investigation, appraisal, and reuse, of a cast iron structural frame". *The structural engineer*, vol 76, no 3, pp 37–42

Cadei, J M C (1998). "Factors of safety in the limit state design of FRP composite structures". In: *Proc designing cost-effective composites, Sep, IMechE, London*. Professional Engineering Publishing, London, pp 27–40

Cadei, J M C and Stratford, T J (2004). "Elastic analysis of adhesion stresses between a beam and a bonded strengthening plate". To be presented at ACIC-2004

Chalkly, P D (1993). *Mathematical modelling of bonded fibre-composite repairs to metals*. Aeronautical Research Laboratory Research Report AR-008-365, Department of Defence and Technology Organisation, Australia

Church, D G and Silva, T M D (2002). "Application of carbon fibre composites at covered ways 12 and 58 and bridge EL". In: *Proc ACIC 2002 – inaug int conf use of advanced composites in construction, 15–17 Apr, Univ of Southampton*. Thomas Telford, London

CIRIA (1977). *Rationalisation of safety and serviceability factors in structural codes*. Report 63, CIRIA, London

Clarke, J L, ed (1996). *Structural design of polymer composites. Eurocomp design code and handbook*. E & FN Spon, London

Collins, R and Conroy, A (2001). "Composites: design for deconstruction, reuse and recycling". In: *Proc NGCC 1st ann conf & AGM, Composites in construction; through life performance, 30–31 Oct, Watford*

<http://www.concrete-repairs.co.uk/news/pr04_bridge0999.htm>

<http://www.concrete-repairs.co.uk/news/pr07_carbon 0300.htm>

Concrete Society (2000). *Design guidance for strengthening concrete structures using fibre composite materials*. Technical Report 55, Concrete Society, Crowthorne

Concrete Society (2003). *Strengthening concrete structures with fibre composite materials: acceptance, inspection and monitoring*. Technical Report 57, Concrete Society, Crowthorne

The Construction (Design and Management) Regulations 1994. SI 1994/3140, HMSO, London

Control of Substances Hazardous to Health Regulations 1999. SI 1999/437, Stationery Office, London

Cripps, A, ed (2001). *Fibre-reinforced polymer composites in construction*. C564, CIRIA, London

Cullimore, M S G (1967). "The fatigue strength of wrought iron after weathering in service". *The structural engineer*, vol 45, no 5, pp 193–199

Dao, M and Asaro, R J (1999). "A study on failure prediction and design criteria for fiber composites under fire degradation". *Composites, Part A*, vol 30, pp 123–131

Davies, J M, Dewhurst, D, McNicholas, J B and Wang, H-B (1995). "Fire resistance by test and calculation". In: *Proc int conf fire safety by design, Univ of Sunderland, Jul*

Demers, C (1998). "Tension-tension axial fatigue of E-glass fiber-reinforced polymeric composites. Fatigue life diagram". *Const and bldg materials*, vol 12, no 1, pp 303–310

Denton, S R (2001). "Analysis of stresses developed in FRP plated beams due to thermal effects". In: J G Teng (ed), *Proc int conf FRP composites in civ engg (CICE 2001), 12–15 Dec, Hong Kong*, pp 527–536

Department of Transport (1994). *Strengthening of concrete highway structures using externally bonded plates*. BA 30/94 (DMRB vol 3, sec 3, pt 1), HMSO, London

DNV (2003) *Composite components*. Offshore Standard DNV-OS-C501, Det Norske Veritas, Oslo

Doran, D K, ed (1992). *Construction materials reference book*. Butterworth Heinemann, London

Farmer, N and Smith, I (2001). "King Street Railway Bridge – strengthening of cast iron girders with FRP composites". In: M C Forde (ed), *Proc 9th int conf struct faults and repair, 4–6 Jul, London*. Engineering Technics Press

fib (2001). *Externally bonded FRP reinforcement for RC structures*. fib bulletin 14, Fédération internationale du béton, Lausanne

Findley, W N (1971). "Combined stress creep of non-linear viscoelastic material". In: A L Smith and A M Nicolson (eds), *Advances in creep design*. Applied Science Publications, London

Frost, S, Lee, R J and Thompson, V K (2003). "Structural integrity of beams strengthened with FRP plates – analysis of the adhesive layer". In: *Proc struct faults and repair 2003, London*

Garden, H N (2001). "Use of composites in civil engineering infrastructure". *Reinforced plastics*, Jul/Aug, vol 45, no 7/8, pp 44–50

Gibson, A G, Wu, Y S, Chandler, H W, Wilcox, J A D and Bettess, P (1995). "A model for the thermal performance of thick composite laminates in hydrocarbon fires, composite materials in the petroleum industry". *Revue de l'Institute Francais du Petrole* (special issue), vol 50, no 1, pp 60–74

Goland, M and Reissner, E (1944). "The stresses in cemented joints". *J app mechanics*, trans ASME, no A17

Griffith, W I, Morris, D H and Brinson, H F (1978). *The accelerated characterisation of composite materials*. VPI engineering series VPI-E-78-3. Virginia Polytechnic Institute College of Engineering, Blacksburg, VA

Hahn, H (1976). "Fatigue behaviour of composite laminates". *J composite materials*, vol 10, pp 156–180 and pp 266–278

Hambly, E C (1991). *Bridge deck behaviour*. 2nd edn, Spon Press, London

Hart-Smith, L J (1973). *Adhesive-bonded double-lap joints*. Contractor Report CR-112235, NASA, Langley, USA

Hart-Smith, L J (1994). "The key to designing durable adhesively bonded joints". *Composites*, vol 25, no 9, pp 895–898

Health and Safety at Work etc Act 1974. 1974 c. 37, HMSO, London

Henderson, J B and Wiecek, T E (1987). "A mathematical model to predict the thermal response of decomposing expanding polymer composites". *J composite materials*, vol 21, pp 373–393

Hertzberg, R W (1996). *Deformation and fracture mechanics of engineering materials*. 4th edn, John Wiley & Sons, pp 201–375 and pp 521–589

Highways Agency (1996). *The assessment of steel highway bridges and structures*. BD 56/96 (DMRB vol 3, sec 4, pt 11), Stationery Office, London

Highways Agency (2001a). *The assessment of highway bridges and structures*. BD 21/01 (DMRB vol 3, sec 4, pt 3), Stationery Office, London

Highways Agency (2001b). *Loads for highway bridges*. BD 37/01 (DMRB vol 1, sec 3, pt 14), Stationery Office, London

Hollaway, L C and Head, P R (2001). *Advanced polymer composites and polymers in the civil infrastructure*. Elsevier Science

Hull, D and Clyne, T W (1996). *An introduction to composite materials*. 2nd edn, Cambridge University Press

Hutchinson, A R (1997). *Joining of fibre-reinforced polymer composite materials*. Project Report 46, CIRIA, London

IStructE (1996). *Appraisal of existing structures*. 2nd edn, SETO, London

IStructE (1999). *A guide to the structural use of adhesives*. SETO, London

Jones, C (1994). "Fatigue of GFRP". *J Composite materials*, vol 28, pp 309–327

Jones, N (2001). "Spray on strength". *New Scientist*, vol 172, no 2313, p 26

Jones, R M (1998). *Mechanics of composite materials*. 2nd edn, Taylor & Francis

Kinloch, A J (1983). *Durability of structural adhesives*. Elsevier Applied Science Pub, Amsterdam, ch 1

Lainchbury, J and Edwards, S (2001). "The life cycle environmental impacts of construction products". In: *Proc NGCC 1st ann conf & AGM, Composites in construction; through life performance, 30–31 Oct, Watford*

Lane, I R and Ward, J A (2000). "Restoring Britain's bridge heritage". Paper presented to Instn Civ Engrs (South Wales Association) Transport Engineering Group

Lee, S M (1990). *International encyclopaedia of composites*. John Wiley & Sons, vol 2, pp 107–111

Leonard, A R (2002). "The design of carbon fibre composite strengthening for cast iron struts at Shadwell Station vent shaft". In: *Proc ACIC 2002, inaug int conf use of advanced composites in construction, 15–17 Apr, Univ of Southampton*. Thomas Telford, London

Liu, X, Silva, P F and Nanni, A (2001). "Rehabilitation of steel bridge members with FRP composite materials". In: *Proc 1st int conf composites in construction (CCC 2001), 10–12 Oct, Porto, Portugal*, pp 613–617

Luke, S (2001a). "Strengthening of structures with carbon fibre plates – case histories for Hythe Bridge, Oxford and Qafco Prill Tower, Qatar". In: *Proc NGCC 1st ann conf & AGM, Composites in construction; through life performance, 30–31 Oct*, Watford

Luke, S (2001b). "Strengthening of existing structures using advanced composite materials". NGCC paper presented at IStructE, 18 Sep 2001

Luke, S (2001c). "The use of carbon fibre plates for the strengthening of two metallic bridges of an historic nature in the UK". In J G Teng (ed), *Proc int conf FRP composites in construction (CICE 2001), 12–15 Dec, Hong Kong*, pp 975–983

Mandell, J. (1982). "Fatigue behaviour in fibre resin composites". In: G Pritchard (ed), *Developments in reinforced plastics*. Kluwer, Dordrecht, vol 2, pp 67–107

Mays, G C and Hutchinson, A R (1992). *Adhesives in civil engineering*. Cambridge University Press

Mertz, D R, Gillespie, J W, Chajes, M J and Sabol, S A (2001). *The rehabilitation of steel bridge girders using advanced composite materials*. IDEA program final report, contract no NCHRP-98-ID051, Transportation Research Board, National Research Council

Miller, T C (2000). "The rehabilitation of steel bridge girders using advanced composite materials". Master's thesis, University of Delaware, Newark

Miller, T C, Chajes, M J, Mertz, D R and Hastings, J N (2001). "Strengthening of a steel bridge girder using CFRP plates". In: *Proc New York City bridge conf, 29–30 Oct, New York*

Morgan, J (1999). "The strength of Victorian wrought iron". *Proc Instn Civ Engrs, Structs & bldgs*, vol 134, Nov, pp 295–300

Moy, S S J (1999). "A theoretical investigation into the benefits of using carbon fibre reinforcement to increase the capacity of initially unloaded and preloaded beams and struts". Paper prepared under Link & Surface Transport Programme, Carbon fibre composites for structural upgrade and life extension – validation and design guidance. Dept of Civil & Environmental Engg, Univ Southampton

Moy, S S J (2001a). "Early age curing under cyclic loading – a further investigation into stiffness development in carbon fibre reinforced steel beams". Paper prepared under Link & Surface Transport Programme, Carbon fibre composites for structural upgrade and life extension – validation and design guidance. Dept of Civil & Environmental Engg, Univ Southampton

Moy, S S J, ed (2001b). *FRP composites – life extension and strengthening of metallic structures*. ICE design and practice guide, Thomas Telford, London

Moy, S S J (2002). "Early age curing under cyclic loading – an investigation into stiffness development in carbon fibre reinforced steel beams". In: *Proc ACIC 2002, inaug int conf use of advanced composites in construction, 15–17 Apr, Univ Southampton*. Thomas Telford, London

Moy, S S J, Barnes, F, Moriarty, J, Dier, A F, Kenchington, A and Iverson B (2000). "Structural upgrade and life extension of cast iron struts and beams using carbon fibre reinforced composites". In: A G Gibson (ed), *Proc 8th int conf fibre reinf composites, FRC 2000 – composites for the millennium, 13–15 Sep, Univ Newcastle-upon-Tyne*

Moy, S S J, Nikoukar F (2002). "Flexural behaviour of steel beams reinforced with carbon fibre reinforced polymer composite". In: *Proc ACIC 2002, inaug int conf use of advanced composites in construction, 15–17 Apr, Univ Southampton*. Thomas Telford, London

New Civil Engineer (2000). "Historic Bridge Award 2000 winners". *New civ engr*, no 1378, 23 Nov, p 26

New Civil Engineer (2002). "Custodian of the lists. Joining the cast". *New civ engr*, no 1429, 14 Feb, p 25

Pierron, F, Poitette, Y and Vautrin, A (2002). "A novel procedure for identification of 3D moisture diffusion parameters on thick composites, theory, validation and experimental results". *J composite materials*, vol 36, no 19, pp 2219–2244

Rajagopalan, G, Immordino, K M and Gillespie, J W (1996). "Adhesive selection methodology for rehabilitation of steel bridges with composite materials". In: *Proc 11th tech conf on composites*. American Society for Composites, Atlanta, p 222

Reifsnider, K L (1982). "Analysis of fatigue damage in composite laminates". *Int j fatigue*, vol 2, no 1, pp 3–11

Roberts, T M (1989). "Approximate analysis of shear and normal stress concentrations in the adhesive layer of plated RC beams". *The structl engr*, vol 10, no 2, pp 229–233

Rotem, A (1993). "Load frequency effect on fatigue strength isotropic laminates". *Composites science and technology*, vol 46, no 2, pp 129–139

Sen, R, Liby, L and Mullins, G (2001). "Strengthening steel bridge sections using CFRP laminates". *Composites, Part B: Engineering*, vol 32, no 4, pp 309–322

Sen, R, Liby, L, Spillett, K and Mullins, G (1995). "Strengthening steel composite bridge members using CFRP laminates". In: L Taerwe (ed), *Proc 2nd int symp non-metallic (FRP) reinf for conc strucs (FRPRCS-2), Aug, Ghent*. E & FN Spon, London, pp 551–558

Shenton, H W, Chajes, M J, Finch, W W, Hemphill, S and Craig, R (2000). "Performance of a historic 19th century wrought iron through-truss bridge rehabilitated using advanced composites". In: *Proc ASCE advanced tech in structural engg: structures congress 2000, 8–10 May, Philadelphia*. ASCE, Reston, VA

Smith, S T and Teng, J G (2001). "Interfacial stresses in plated beams". Engineering structures, vol 23, pp 857–871

Stöcklin, I and Meier, U (2001). "Strengthening of concrete structures with prestressed and gradually anchored CFRP strips". In: *Proc 5th int conf FRP reinf conc strucs (FRPRCS-5), 16–18 Jul, Cambridge*. Thomas Telford, London, pp 291–296

Swailes, T (1995). "19th century cast iron beams: their design, manufacture and reliability". *Proc Instn Civ Engrs, Civ engg*, vol 114, Feb, pp 25–35

Swailes, T and Marsh, J (1998). *Structural appraisal of iron-framed mills*. ICE design and practice guides, Thomas Telford, London

<http://www.tgp.co.uk/feature/frp4/kingstreet.html>

Tucker, W C, Brown, R and Russell, L (1990), "Corrosion between a graphite/polymer composite and metals". *J composite materials*, vol 24, no 1, pp 92–102

Tuttle, M E, Mescher, A M and Potocki, M L (1997). "Mechanics of polymeric composites exposed to a constant heat flux". In: *Proc ASME intl mech engg conf: Composites and functionally graded materials*. MD-Vol 80, ASME, Dallas

West, T D (2001). *Enhancement to the bond between advanced composite materials and steel for bridge rehabilitation*. CCM Report 2001-04, University of Delaware Center for Composite Materials

BRITISH AND OTHER STANDARDS

BS

BS 476-7:1997. *Fire tests on building materials and structures. Method of test to determine the classification of the surface spread of flame of products*

BS 476-11 to 20:1987. *Fire tests on building materials and structures. Method for determination of the fire resistance of elements of construction* [various subtitles]

BS 1452:1990. *Specification for flake graphite cast iron* [superseded]

BS 1881-207:1992. *Testing concrete. Recommendations for the assessment of concrete strength by near-to-surface tests*

BS 2782-3:Methods 320A to 320F:1976. *Methods of testing plastics. Mechanical properties. Tensile strength, elongation and elastic modulus* [obsolescent]

BS 4994:1987. *Specification for design and construction of vessels and tanks in reinforced plastics*

BS 5350-C5:2002. *Methods of test for adhesives. Determination of bond strength in longitudinal shear for rigid adherends*

BS 5400-1:1988. *Steel, concrete and composite bridges. General statement*

BS 5400-10:1980. *Steel, concrete and composite bridges. Code of practice for fatigue*

BS 5950:2000. *Structural use of steelwork in building. Code of practice for design. Rolled and welded sections*

BS 6319-3:1990. *Testing of resin and polymer/cement compositions for use in construction. Methods for measurement of modulus of elasticity in flexure and flexural strength*

BS 6319-8:1984. *Testing of resin and polymer/cement compositions for use in construction. Method for the assessment of resistance to liquids*

BS 7079-0:1990. *Preparation of steel substrates before application of paints and related products. Introduction*

BS 7608:1993. *Code of practice for fatigue design and assessment of steel structures*

BS 7991:2001. *Determination of the mode I adhesive fracture energy, GIC, of structural adhesives using the double cantilever beam (DCB) and tapered double cantilever beam (TDCB) specimens*

BS 8110-1:1997. *Structural use of concrete. Code of practice for design and construction*

BS EN

BS EN 1561:1997. *Founding. Grey cast irons*

BS EN 1563:1997. *Founding. Spheroidal graphite cast iron*

BS EN 1770:1998. *Products and systems for the protection and repair of concrete structures. Test methods. Determination of the coefficient of thermal expansion*

BS EN 1990:2002. *Eurocode. Basis of structural design*

BS EN 2378:1994. *Fibre reinforced plastics. Determination of water absorption by immersion*

BS EN 14022:2003 Structural adhesives. *Determination of the pot life (working life) of multicomponent adhesives*

BS EN ISO

BS EN ISO 527-4:1997/BS 2782-3:Method 326F:1997. *Plastics. Determination of tensile properties. Test conditions for isotropic and orthotropic fibre-reinforced plastic composites*

BS EN ISO 527-5:1997/BS 2782-3:Method 326G:1997. *Plastics. Determination of tensile properties. Test conditions for unidirectional fibre-reinforced plastic composites*

BS EN ISO 899-1:2003. *Plastics. Determination of creep behaviour. Tensile creep*

BS EN ISO 899-2:2003. *Plastics. Determination of creep behaviour. Flexural creep by three-point loading*

BS EN ISO 4624:2003/BS 3900-E10:2003. *Paints and varnishes. Pull-off test*

BS EN ISO 9002:1994. *Quality systems. Model for quality assurance in production, installation and servicing*

BS EN ISO 14125:1998. *Fibre-reinforced plastic composites. Determination of flexural properties*

BS EN ISO 14126:1999. *Fibre-reinforced plastic composites. Determination of compressive properties in the in-plane direction*

BS EN ISO 14129:1998. *Fibre-reinforced plastic composites. Determination of the in-plane shear stress/shear strain response, including the in-plane shear modulus and strength by the ± 45° tension test method*

BS EN ISO 14130:1998. *Fibre-reinforced plastic composites. Determination of apparent interlaminar shear strength by short-beam method*

BS ISO

BS ISO 1268-1:2001 to 1268-9:2003. *Fibre-reinforced plastics. Methods of producing test plates* [various subtitles]

BS ISO 11357-2:1999. *Plastics. Differential scanning calorimetry (DSC). Determination of glass transition temperature*

BS ISO 11359-2:1999. *Plastics. Thermomechanical analysis (TMA). Determination of coefficient of linear thermal expansion and glass transition temperature*

BS ISO 15310:1999. *Reinforced plastics. Determination of the in-plane shear modulus by the plate twist method*

ISO

ISO 6721-1:2001. *Plastics. Determination of dynamic mechanical properties. General principles*

ASTM

ASTM E 1640-99. *Standard test method for assignment of the glass transition temperature by dynamic mechanical analysis*

Strengthening metallic structures using externally bonded fibre-reinforced polymers

J M C Cadei

T J Stratford

L C Hollaway

W G Duckett

CIRIA *sharing knowledge ▪ building best practice*

Classic House, 174–180 Old Street, London EC1V 9BP, UK
TEL +44 (0)20 7549 3300   FAX +44 (0)20 7253 0523
EMAIL enquiries@ciria.org
WEBSITE www.ciria.org

Summary

This book provides independent best practice guidance on the design, selection, detailing, installation and subsequent maintenance of externally bonded fibre-reinforced polymers for metallic structures. It provides advice on the most appropriate methods, discussing the advantages and disadvantages of each method, both in terms of cost-effectiveness and long-term performance.

The guidance is aimed at those who design the strengthening works as well as those who carry out the installation. It is also relevant in terms of broader asset management considerations.

Strengthening metallic structures using externally bonded fibre-reinforced polymers

Cadei, J M C; Stratford, T J; Hollaway, L C; Duckett, W G

CIRIA

Publication C595   © CIRIA 2004   ISBN 0-86017-595-2   RP645

Keywords		
Materials technology, project management, QA, refurbishment, transport infrastructure		

Reader interest	Classification	
Asset managers, maintenance and inspection engineers, designers and installers of strengthening works, managers and operators of structures	AVAILABILITY	Unrestricted
	CONTENT	Advice/guidance
	STATUS	Committee-guided
	USER	Structural engineers, designers and contractors, asset managers

Published by CIRIA, Classic House, 174–180 Old Street, London EC1V 9BP, UK.

British Library Cataloguing in Publication Data

A catalogue record is available for this book from the British Library.